SHEDDING THE VEIL

SHEDDING THE VEIL

MAPPING THE EUROPEAN DISCOVERY OF AMERICA AND THE WORLD

---

based on selected works from

THE SIDNEY R. KNAFEL COLLECTION OF EARLY MAPS, ATLASES, AND GLOBES

1434 - 1865

THOMAS SUÁREZ

World Scientific
Singapore • New Jersey • London • Hong Kong

The Gallery
Federal Reserve Board
Washington, D.C.
October 1 - November 30, 1991

———————

The Addison Gallery of American Art
Phillips Academy
Andover, Massachusetts
April 10 - June 14th, 1992

———————

Bristol-Myers Squibb Gallery
Princeton, New Jersey
October 4 - November 29, 1992

———————

Published by

WORLD SCIENTIFIC PUBLISHING CO. PTE. LTD.
P. O. Box 128, Farrer Road, Singapore 9128
USA office: Suite 1B, 1060 Main Street, River Edge, NJ 07661
UK office: 73 Lynton Mead, Totteridge, London N20 8DH

Library of Congress Cataloging-in-Publication Data

Suárez, Thomas.
    Shedding the veil : mapping the European discovery of America and
the world / Thomas Suárez.
        p. cm.
    Based on selected works from the Sidney R. Knafel Collection of
early maps, atlases, and globes. 1434-1865.
    ISBN 9810208693
    1. America — Discovery and exploration — Maps.
2. Geography — 15th - 16th centuries. 3. Cartography — History.
I. Title.
E121.S87   1992
970.01 – dc20   92-6466
                                                                              CIP

Printed in Singapore.

For my wife Ahngsana . . .

ACKNOWLEDGEMENTS

The Sidney R. Knafel Collection

The author is indebted to Sidney Knafel for the opportunity to research selected maps from his collection, and to use them as the vehicle for this project. Without his substantial help and kindness, this book would not have been possible.

Author's Thanks

I would like to thank all those who assisted me in the preparation of this work. In particular, I am extremely grateful to Richard Casten for his scrutiny of my draft, for his ideas, inspirations, untiring encouragement, and most of all his friendship. I am also particularly indebted to David Woodward for his kind, generous, and meticulous scholarly guidance, and for rescuing me from several would-be pitfalls. Don McGuirk and Arthur Holzheimer were invaluable for their excellent suggestions, insights, and source material. William Haver led me through Asian matters, and Dudley Barnes helped with Batista Beccari. William Haver copy-edited the draft. John Suárez proofread the final copy.

Yvonne Yip at World Scientific Publishing (Singapore) selflessly lent her impressive expertise and talents, caring for the production of the book as though it were her own. I am truly grateful to her and to the other people at World Scientific who assisted, all of whom made me feel especially fortunate in having World Scientific as my publisher.

My daughter Sainatee helped keep life in perspective. She eternally cheered me up, and would yank me back from the sixteenth century when I would get perilously disoriented there. Apparently fullfilling a sense of filial piety, she dutifully made certain that she (and, therefore, I) always listened to Ravel's *l'Enfant et les Sortilèges* in my office while I worked (be cautioned that cats and various nocturnal forest creatures may have escaped into these pages from sections X and XI of that work).

Above all, I would like to thank Ahngsana for keeping me from falling into the Ocean Sea and off the edge of the Earth while writing this book, and to thank my parents for everything imaginable.

Photography Credit

The photography for this book (except plate 32) was done by Charlotte Raymond, with funding generously provided by the Bristol-Myers Squibb Company.

Thomas Suárez, New York, September, 1991

CONTENTS

LIST OF MAPS

A COLLECTOR'S COMMENT

The fifteenth century simultaneously produced surges in the exploration of humanity, ideas, and science. Just as book production facilitated the assertion of new beliefs and the expansion of knowledge, map and atlas production was an integral part of the exploration of the world. Maps recorded geographical discovery and gave a report to the public. They became the travel guide for the next probe into the unknown.

Collecting maps to uncover the history of exploration involves more than locating the links of a chain, however. Knowledge of the world developed unevenly; misconceptions were superseded by other misconceptions nearly as often as by the accurate observations that eventually prevailed. Map collecting, then, offers the frustration of working on a three-dimensional puzzle and the satisfaction of always discovering more about discovery. That these records can be decorative and on occasion amusing provides an extra reward for this collector.

My collection, let alone this book, would not have been possible without the selfless guidance provided by Thomas Suárez. Bearing the splendid traits which characterize very nearly all those who concern themselves commercially with cartography, he has taught what he learned, provided what he obtained, and encouraged what he feared. His grasp of the events which were revealed by maps has served me well and, now with this book, serves all of us a broad and interesting story.

Sidney R. Knafel

A NOTE ABOUT THE MAPS AND TEXT

The maps, atlases, portolan charts, and globe chosen from the Collection for this volume were selected to show the European discovery of the world, the history of the world's lifting its veil to explorers, mapmakers, and ultimately the people, beginning with the cartographic depiction of the world before Columbus and developing through American coastal charts of the nineteenth century. Its intent is neither to glorify nor condemn the traumatic events of the past five centuries; its perspective is necessarily Eurocentric because it is seen through the eyes of her maps as interpreted by this particular Western observer.

The progression of maps reflects an evolving concept of the earth, the political claims which led to modern world divisions, and the art and science of mapmaking itself. An emphasis has been placed on the earliest material. Maps which would merely repeat a point have been avoided, and an effort has been made to show maps of all esthetic modes. The following cautions, however, should be noted :

All claims to primacy, whether of an explorer's accomplishments or a map's geography, should be taken only in context of our very fragmented knowledge of the past. It is a given that the early voyages known to us represent only a random sampling of those that actually took place; undoubtedly, there have been monumental exploratory voyages of which no knowledge survives. Similarly, the early maps upon which we trace cartographic evolution are restricted to those which, owing either to their production in sufficiently high quantity or to fortune of circumstance, have beat the very poor odds of their survival to our time. If both these considerations seem fairly obvious, they have also commonly gone unheeded.

The dating of a work refers to the year it was published or made. After 1500 this is to varying degrees contemporary with the geographic and cosmographic thinking of the time as dictated by current explorations and scientific theories. Although through chapter 3 the maps depict concepts which originated at a far earlier time, even those incunabula presented what was "current knowledge" to the literate public who studied them. When a second date appears in parentheses, it refers to the actual date of issue, while the first refers to the original year the map was published. Some works have been put slightly out of strict chronological order of publication to show more logically a sequence or evolution.

T. S.

PART ONE — A WORLD CALLED EUROPE

Chapter I

PRECEDENTS

AS EARLY AS THE THIRTEENTH CENTURY a new humanistic attitude began to permeate the European psyche, one which dared to put trust in the worthiness of the individual. This was the dawn of the Renaissance, the beginning of the modern world. Its energy affected all aspects of society: arts and science, social and political character, the way people viewed the outside world and the place they perceived themselves occupying on earth and in the cosmos. Although this chapter in history has long ended, its philosophies are still a prevailing force in the world today.

By the fifteenth century change had become irreversible. The feudal system was being consumed by larger social structures with centralized governments, and resources could be better pooled toward a chosen goal. Science emerged as a pursuit increasingly independent of theological and philosophical constraints. Artists painted subjects which exuded emotion rather than containing it, creating art which exposed the body, ridiculed the State, and even mocked the Church. Secular *ars nova* music flourished. Classical knowledge was sought and venerated. Despite the reality of Church power and contemporary politics, as manifested in such formidable obstacles as inquisitions and the need to please one's patron, somehow creativity flourished. It was in this context that an eccentric dreamer, known as Cristoforo Colombo in his native Genoa and Cristobal Colón in Spain, persevered in an obsession with crossing the western ocean. His energy and enterprise expedited the most dramatic and inevitable expression of the Renaissance : the recorded Discovery of the World.

Five centuries later, Columbus has become more of a symbol than a hero.[1] That he was not the first European to reach American shores has been amply demonstrated; that he was deceitful, greedy, opportunistic, and mis-managed his colonial enterprises is clear; that he treated his indigenous hosts with savagery and contempt is evident from contemporary records. Perhaps even Columbus' courage has been overstated : we often conjure an image of the Genoese explorer steadfastly heading west into a mysterious sea towards some purported end of the earth. But few people then believed the earth to be flat, and the precedent to head west into the Atlantic had already been well established by the Portuguese and others.

Columbus' destiny in exploration was in fact closely entwined with that of Portugal, even though he ultimately sailed under the Castilian flag. Under a brilliant blueprint of scrupulous and methodical exploratory policies formulated by Prince Henry "the Navigator" in the early fifteenth century, Portugal simultaneously pursued two fronts, pushing both southward along the African coast and westward into the Atlantic. The African venture clearly held the Orient as its goal; the objective of the Atlantic initiative was more abstract but soon

---

1. Recent scholarship has made strikingly clear how little is factually known about Columbus. For a critical appraisal of the man and his legacy see e.g., Carl Ortwin Sauer, The Early Spanish Main, and Kirkpatrick Sale, The Conquest of Paradise.

resulted in the discovery and colonization of the Azores, a significant stepping-stone to the discovery of America.[2]

The circum-African scheme dominated, however. But in 1474 King Afonso V, finding progress along the African coast to be slow and unfruitful, considered giving higher priority to the westward option as a way to the Orient. On this matter he solicited the opinion of the most highly esteemed cosmographer known to him, Paolo Toscanelli of Florence. In a now-famous letter sent in June of that year, Toscanelli argued that a trans-Atlantic route to Asia would be preferable to the African route. He vastly undersized the globe and placed intermediate islands as ports-of-call, thus minimizing the logistical problems of such a crossing.[3]

Portuguese scientists appear, correctly, to have doubted the diminutive figure for the earth's circumference proposed by Toscanelli (and later by Columbus), and thereby doubted the feasibility of such a crossing to the Orient, but certainly not the trans-Atlantic concept itself. Private auspices continued to fund excursions into the western ocean, with the Azores serving as an ideal mid-Atlantic outpost.

Although the winds west of the Azores are unfavorable for an outbound ocean crossing, some adventurers did in fact claim to have reached land far west of their Azorian base. Columbus, aware of these tales, frequented Portuguese ports hunting news and rumors of such voyages.[4] He lived and married in Madeira, another recent Portuguese outpost, which had become an active exporter of wine and sugar, and thus was a likely source of exploratory gossip.[5] Perhaps most compelling is Columbus' early contacts with the Canaries, the islands from which he ultimately launched his trans-Atlantic voyages. The Canaries are ideally situated for the west-bound journey because of their favorable winds; it remains unknown whether Columbus had gathered information to that effect from other pilots, guessed it on his own, or whether his choice of the Spanish-governed Canaries was politically motivated.[6]

---

2. See footnote 67 for Azores discovery.

3. Africa proved to extend much further south than expected. The southern shores of West Africa at first appeared to be the "bottom" of the continent, and so its subsequent and seemingly interminable turn to the south was unexpected. Ironically, the belief that the southern coast of West Africa marked the end of the continent was reinforced by the fact that it lies at about the latitude where some early maps, e.g., that of Macrobius (see entry 3), suggested the continent would end. The shore's subsequent turn to the south was an unwelcome surprise for Portugal, and Afonso V was unsure whether to put more effort into the African or Atlantic route to Asia. Additionally, a private contract for Portugal's African enterprise expired in 1474, probably accounting for the particular year Toscanelli was approached. Toscanelli expressed the opinion that the Indies could be reached by sailing west, and that this route was in fact preferable to the African route.

4. Speculative arguments have been advanced in favor of a pre-Columbian Portuguese landfall in America. See pages 33-34 in entry 9 (Beccari).

5. The Madeira group had been occasionally sighted by Spanish and Portuguese sailors returning from the Canaries, but was formally "discovered" and claimed by Portugal in 1419. Colonization began after the principal island, which was completely forested, was set to flames by Portuguese, supposedly burning for seven years.

6. Columbus' choice of the Canaries as the launching point for his ocean crossing has aroused much speculation, as it would seem to have been an unlikely choice for an Iberian pilot attempting to reach Asia via the shortest route possible. One can only wonder whether Columbus would have chosen the Azores rather than the Canaries for the point of departure had his benefactor been the Portuguese Crown. Yet of all Columbus' options, it was only from the Canaries that prevailing winds would have enabled his success. He also correctly chose a more northerly latitude for the return crossing. His decision may have been the result of uncanny luck, brilliant deduction, or unadvertised knowledge gained through the reports of other voyages. See also footnote 72.

In 1484 Columbus, then a resident of Lisbon, approached Afonso V's successor, John II, to entreat backing for his proposed voyage into the western ocean. He may have been turned away not, as posterity tells us, because the Crown's advisors dismissed his plan as madness, but rather because they had already been engaged in expeditions to the west for at least a half century. The proposal Columbus presented to John II was essentially the same espoused to the Portuguese Crown by Toscanelli a decade earlier, a proposal which Columbus himself had solicited and read in the interim.[7] Columbus was seen to be a late-comer, merely following in the path they had already forged. Portuguese interest in a western route to the Indies did not wane until 1488, when the survivors of Bartolomeo Dias' fleet announced that they had entered the Indian Ocean by rounding the southern tip of Africa. By early 1492, Spain saw Portugal's costly gambles in exploration beginning to pay off; it would have seemed a folly for Ferdinand and Isabella to miss an opportunity to compete. Further, they had little to lose: the offer that Columbus presented to them was "back-weighted," reserving an avaricious reward for himself and his heirs only in the event of his success. Freshly rid of Moorish clutches and perhaps growing envious of Portugal's seemingly imminent mastery of a sea route to the East, Spain was ready to flex her own muscles. Thus as a lever in the prying open of the world, Columbus, beyond his genius of perseverance and the astuteness of his choice of sailing routes, was simply a product of his times.

Columbus' Landfall in the Caribbean

En route across the Atlantic on Thursday, October 11, 1492, at about 10:00 at night, Christopher Columbus thought he saw a light or torch in the distance to the west.[8] He had become increasingly nervous about reaching land soon, and so he doubled the number of people keeping watch and reminded his men that the first to sight land would be awarded a silk doublet from him personally, and an annuity of 10,000 *maravedis* from the Crown. At roughly two hours past midnight one Juan Bermejo from the *Pinta* yelled *"Tierra"* and the vessel fired its cannon, signal for the sighting of land. America had formally been "discovered." Columbus, in retrospect, believed that the faint light he thought he had seen earlier had been a premonition from God (and so he claimed the reward for himself).[9]

At dawn Columbus and some of his men went ashore, planted flags, and proclaimed possession of the land for Ferdinand and Isabella. Columbus states in his log that the people of the island were handsome (save for overly broad heads and foreheads), friendly, and naked. They called their land *Guanahani*. Columbus dubbed it *San Salvador*.

The identification of *Guanahani*, though perhaps not really an important point in the whole scheme of things, has long been an issue of impassioned contention. Candidates have included the islands Cat, Conception, Plana Cays, Mayaguana, East Caicos, Grand Turk, Watling, and Samana Cay. In modern times Watling Island (now called San Salvador) has been the preferred theory, being supported by Becher (1856), Murdoch (1884), Thacher (1896), and, most influentially, Morison (1942). A recent project by *The National Geographic Society* concluded that Samana Cay was the site of Columbus' maiden landing, supporting the

---

7. Columbus read a copy of Toscanelli's proposal of an Atlantic crossing to the Orient about 1480, possibly having solicited it from Toscanelli himself. But whether or not Toscanelli actually directly corresponded with Columbus has been challenged; no such original letter survives. The effect Toscanelli had on the course of history has perhaps been traditionally overstated, as the information it contained was not exclusive, though in any event it certainly reinforced and encouraged Columbus' plans.

8. We record the dates according to the Julian calendar then in use. But the day of landfall, October 12 of Columbus' log, actually corresponds to October 23 of our modern Gregorian calendar, which was devised by scientists under Pope Gregory XIII as directed by a papal mandate of 1582.

9. The Crown is believed to have financed the reward from a special tax on Sevillian butchers and/or the confiscation of the property of "conversos," Jews who had "converted" and thus were allowed to remain in Spain.

conclusion reached by Gustavus Fox in 1882.[10] Other scholars, however, still support Watling Island as the initial landfall.[11] In the final analysis, there can never be certainty on the issue, and any pretense of such is naive at best. Even if the methods used to reconstruct the voyage were "perfect" (National Geographic used computers), we have only imperfect records of the voyage and landing from which to work.[12]

The role of maps at the close of the fifteenth century

The fifteenth century was a particularly prosperous time for the city of Mainz, a German port city on the Rhine River which had benefitted commercially by having long been an oasis for Europe's Jews. When Columbus was born, experimenters in Mainz were busy developing what is arguably the single most significant invention in our civilization : moveable type and the printing press.[13] The advent of the printed book by 1454 was both a result of, and a catalyst for, Renaissance thought. It unleashed the power of ideas and information, and blurred the line separating the privileged elite from the uneducated masses. With information and ideas being shared and spread substantially faster and wider than ever before, Europe was beginning to shed her veil of medieval provinciality. Columbus himself was nurtured by the new environment of printing, largely formulating his ideas from printed editions of the writings of Claudius Ptolemy, Marco Polo, and Pierre d'Ailly.

The birth of the printed map followed the introduction of the printed book by less than two decades. Many of the earliest printed maps, however, were stubborn vestiges of the Middle Ages. These maps, copied and re-copied in manuscript until being produced in printed form, show concepts of the Earth which had stagnated in Europe for centuries. They do not show the best geographic knowledge then available. Rather, they reveal the world as known in medieval times, and as still known to most Europeans at the end of the fifteenth century. Though archaic, the dissemination of such maps by means of the printed book was in itself a triumph. Within five years of the debut of these most primitive printed maps, an entirely different sort of map, based on classical sources and unknown to medieval Europe, was prepared in printed form.

The ancient Greek and Roman civilizations reached a far higher understanding of the earth and cosmos than was prevalent in the first one and a half millennia of Christian civilization. Our record of classical knowledge is, however, extremely fragmented. The bulk of what we know comes second- and third- hand from the works of people who were more

---

10. Samana Cay is now uninhabited, but artifacts discovered there demonstrate that it was inhabited 500 years ago. See Joseph Judge and James L. Stanfield, "The Island of Landfall" in National Geographic, November, 1986.

11. See, for example, Kim Gainer, "The Cartographic Evidence for the Columbus Landfall," Terrae Incognitae, Vol. XX, p. 69-88 (1988); James Kelley Jr., "The Map of the Bahamas implied by Chaves's 'Derrotero': What is its relevance to the first landfall question?" Imago Mundi vol. 42, p. 26-49 (1990).

12. It was in Columbus' spirit to contort information to suit his conclusions; indeed, he otherwise would never have convinced the Spanish Crown to finance his voyage. His accounting of events is subject to his own errors of judgement, interjections, and romanticization of reality, of which he bore his human share. Further, the original log of the first voyage is not known. The best that we have is a copy by the early historian Bartolomé de las Casas, itself made not from the original but rather from a copy of it prepared by a scribe of Isabella.

13. Printing from moveable type was accomplished far earlier in the Orient. In China, moveable type of baked clay is believed to date back to the Ch'ing-li era (A.D. 1041-49) during the Sung Dynasty, with type made from tin, wood, copper, and lead following later. The Korean language was more suitable for moveable type, and despite the widespread use of Chinese by educated Koreans this may have contributed to the early development of moveable type in Korea. Printing from moveable type in Korea probably dates back to the early thirteenth century (see David Diringer, The Book Before Printing, p. 410-425). The first dated European product of moveable type is a papal indulgence of Gutenberg, 1454.

compilers than innovators. The most influential of these were Strabo and Ptolemy, both of whom lived in Alexandria, where the compilation of knowledge in the classical world reached its culmination.

Alexandria was founded by Alexander the Great upon his "liberation" of Egypt from Persia in 332 B.C. Exploiting the city's strategic position, a series of canals was constructed between Alexandria and the Gulf of Suez to provide easier passage between the Mediterranean and Red Seas (and Indian Ocean). Growing rich from both commerce and the tolls levied for use of the canals, Alexandria evolved into a fabulous and beautiful city.

The most famous of its virtues were its museum and library. By expending the city's enormous resources, scouts were sent all over the known world to acquire books at whatever cost necessary. One ruler, Ptolemy Euergetes, is even said to have had transients searched for books, confiscating any found and returning to them only a copy. Through the very early Christian era, Alexandria flourished as a learned and wealthy cosmopolitan metropolis.

Alexandria was not, however, the only intellectual oasis in the classical world, and her supremacy declined during the last two centuries of the pre-Christian era. During this period, for example, Crates of Mallos worked in Turkey, formulating principles which lacked geographic refinement but presented an intelligent, coherent, and to an extent prophetic concept of the earth, despite its flaws. His ideas were resurrected in the early Christian era by the Roman geographer Macrobius, in which form they provided medieval Europe a breath of classical thought while the writings of Ptolemy were still unknown. Alexandria's gradual period of decline lasted six centuries. In 47 B.C., about two hundred years before Ptolemy, Julius Caesar blockaded Alexandria and books were brought to cloisters surrounding the Temple of Serapis. Most of the books were destroyed when Christian barbarians under Theodosus I sacked the temple and other pagan structures in 391 A.D.

When Europe re-discovered classical knowledge in the Renaissance, other valuable, independent traditions had already established themselves, though not via the printed medium. The portolan chart, for example, offered a far more accurate picture of European shores, and by the later fifteenth century specialized *mappaemundi* recorded parts unknown to the ancient Greeks. Empirical data was applied to the printed map as early as 1475, arguably constituting the first "true" printed map.[14] Classical disciplines, however, still provided the most comprehensive cosmographic overview and the most methodical approach to cartographic theory and science.

- - - - - - - - - - - - - -

---

14. A map of Palestine included in the 1475 Lubeck <u>Rudimentum Novitiorum</u> is the first known European printed map to have been based on actual first-hand observation.

Chapter II

THE WORLD AS KNOWN TO MEDIEVAL EUROPE

1. THE WORLD

[untitled].

Gunther Zainer, Augsburg, 19 November, 1472. Circular schematic map of the world surrounded by a perimeter ocean, in the book of Isidorus Hispalensis (Isidore of Seville), *Etymologiae sive originum libri XX.*

Medium : woodcut.
Size of original : 65 mm diameter.

PLATE : 1

Columbus was a young man of 21 when this rudimentary woodcut map of the world was published. Though modest, it was the first expression of geographic knowledge disseminated by the new medium of printing.[15]

Isidorus of Seville, or Isidore "the Spaniard," was bishop of Seville from circa 599 A.D. until his death in 636. An avid student of both Christian and Roman souces, his *Etymologiae* (between 622-633) was an indiscriminate and uncritical compilation of diverse texts which Isidore himself sometimes did not appear to understand. It was perhaps because of the book's ambiguities and contradictions, rather than in spite of them, that manuscripts of the work maintained a wide appeal and remained a standard encyclopedic text throughout the Middle Ages.

Isidorus' map divides the earth into three realms, those of Asia, Africa, and Europe, by the placement of a "T" shape within a circle, and hence has earned the term "T-O" map.[16] East is at the top. The vertical line of the "T" represented the Mediterranean Sea; as for the horizontal line, the left half traditionally represented the Don River (and the Sea of Azov, Black Sea, Sea of Marmara, and the Aegean Sea), while the right half represented the Nile. There was however great variation among such maps, and this 1472 version simply groups all the waters of the "T" as *Mare magnum sive mediterraneum* ("Great Sea or Mediterranean"). The "Ocean Sea" *(MARE OCEANVM)* surrounds the whole of the Earth in accordance with medieval (and ancient) thought.[17] Noah's three sons, the Biblical progenitors of post-deluge humanity, are indicated in their respective continents: *Sem* in Asia, *Cham* in Africa, and *Japhet* in Europe. Outside the circle the four basic directions are marked; *Oriens* and *Occidens* are of course East and West (for the rising and setting sun),

---

15. This 1472 work of Isidorus is the first printed map of European origin of certain date, and the first in a printed book. The printing of maps appeared earlier in the Orient: e.g., maps printed from woodblocks have been dated to 1155 in China, and 1460 in Korea.

16. The "T" of the T-O map may have been derived from the Greek "tau," an ancient form of the cross which was adopted by early Christians as a clandestine symbol of their faith, symbolically used in the map as superimposing a Christian framework over the entire earth. See Jonathan Lanman, Glimpses of History from Old Maps, ch. 6.

17. The evolution of the concept of the surrounding "ocean sea" can be found on some non-schematic late medieval maps in the circumnavigability of Africa, and by the Renaissance in the circumnavigability of the globe (e.g., entries 12 and 13).

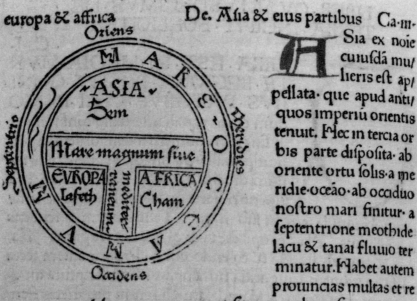

Sia ex noie cuiusdā mu/
lieris est ap/
pellata· que apud anti/
quos imperiū orientis
tenuit. Hec in tercia or
bis parte disposita· ab
oriente ortu solis·a me
ridie·oceāo· ab occiduo
nostro mari finitur· a
septentrione meothide
lacu & tanai fluuio ter
minatur.Habet autem
prouincias multas et re
giones·quarū breuiter nomina et situs expediam·sumpto initio
a paradiso   Paradisus est locus in orientis partibus constitu/
tus·cuius vocabulum ex greco in latinum vertitur ortus.Porro
hebraice eden dicitur·quod in nostra lingua delicie interpretat·
quod vtrumq; iunctum facit ortum deliciarum·est enim omni
genere ligni & pomiferarum arborum consitus habens· etiam
lignum vite.Non ibi frigus· non estus· sed perpetua aeris tem/
peries·e cuius medio fons prorumpens·totum nemus irrigat· di
uiditurq; in quatuor nascentia flumina.Cuius loci post pecca/
tum hominis aditus interclusus est. Septus est eni vndiq; rom
phea flammea·id est muro igneo accinctus· ita ut eius cū caelo
pene iungatur incendium. Cherubin quoq; id est angelorum
presidium arcendis spiritibus malis super romphee flagrantiā
ordinatum est·ut homines flamme·angelos vero malos angeli
boni submoueāt·ne cui carni vel spiritui transgressionis aditus
paradisi pateat.   India vocata ab indo flumine· quo ex parte
occidentali clauditur.Haec a meridiano mari porrecta vsq; ad
ortum solis· & a septentrione vsq; ad montem caucasum perue/
nit·habens gentes multas & oppida· insulam quoq; taprobane
gemmis & elephantibus refertam. Crisam & argiram auro ar/
gentoq; fecundas·vtilem quoq; arboribus foliis nunqm caren
tibus.Habet & flumina gangen & nidan & idaspen illustran/
tes indos. Terra indie fauonio spiritu saluberrima. In anno bis

PLATE 1   Isidore of Seville, Augsburg, 1472   (entry 1)

North is *Septentrio* (the number seven, for the seven plow-oxen stars of the Great or Little Bear) and South is *Meridies* (for the position of the sun at midday).

There is disagreement as to whether the T-O map as a species represented a flat or spherical earth, and indeed it probably represented either according to the prejudices of the viewer. The medieval propensity to believe the earth was flat has, however, traditionally been far overstated, and Isidore, despite the *Etymologiae's* contradictions, makes fairly clear that the earth is spherical.[18] This obviously supported Columbus' cause, although skepticism about the sphericity of the earth was not likely a major obstacle to winning royal support.

Isidore accepted the common premise that Paradise lay in easternmost Asia, Columbus' destination. The *Etymologiae* states that

> *"the Lord planted a garden Eastward in Eden . . . and he placed at the*
> *East of the Garden of Eden, Cherubims, and a flaming sword, which turned*
> *every way, to keep the way of the Tree of Life."* [19]

While a member of Columbus' crew typically might have feared stumbling across this sacred ground, Columbus seemed to relish in the possibility of locating it. Reaching the South American continent on his third voyage, he actually conceded that they were near *"the Earthly Paradise . . . where no man may go, save for by the grace of God."*

It has been demonstrated that Columbus' ultimate goal was to reclaim the Holy Land for Christendom. He grew to perceive himself as one with a divine destiny in the preparation for the coming of the Antichrist and the end of the world, events which he believed were not too distant. For example, in 1500 Columbus wrote that God had made him the messenger of the *"nuevo cielo y mundo"* (i.e., the "new heaven and earth") predicted in the Apocalypse, and that God had shown him where to find it.[20] His log abounds in reflections of divine purpose.

Isidorus' map represents the most extreme divergence from our own twentieth-century Western concept of a "map." The final challenge in understanding it would be for us to see it as a complete map in its own right, not the esoteric schematic relic which it first appears to us to be.

- - - - - - - - - - - - - -

---

18. Although his technical understanding of its principles was sometimes less sophisticated than his Greek sources, e.g., he tried to apply the classical idea of climate zones as non-circumscribing circles lying "flat" on a portion of the earth rather than rings around the earth.

19. Translation from Brown, The Story of Maps.

20. See Pauline Moffitt Watts, "Prophecy and Discovery: On the Spiritual Origins of Christopher Columbus' 'Enterprise of the Indies'," American Historical Review, vol. 90, February, 1985.

2. THE WORLD

[untitled].

[anonymous] Pierre le Rouge, Paris, 1488. Circular *mappamundi* from the book *Mer des Hystoires* (after the *Rudimentun Novitiorum* map of Lubeck, 1475).

Medium : woodcut, with original hand color.
Size of original : 360 mm diameter.

COLOR PLATE : I

The next map to appear in a printed book was geographically typical of the lavish, encyclopedic *mappaemundi* which often hung in medieval monasteries or palaces. The three principal lands of the world —— Asia, Africa, Europe —— are laid out in the same tripartite skeleton as the T-O map, but now with specific countries and regions of the world recorded. Each is symbolized by a mountain, and partitioned by arbitrary waterways. The "T" itself of the T-O pattern is not drawn, and the waters it represented (Mediterranean Sea, Don and Nile Rivers) are not otherwise shown. The map itself would not contradict a flat earth if that is how its viewer chose to see it.

Although vastly more sophisticated than the T-O map, this medieval map utterly fails to record the finest geographic data then available. This, however, was not necessarily the goal of its author. As with the T-O map, to judge it too harshly on these terms would be to impose current values on a world immeasurably different from our own.

At the top, in the farthest region of the East, we see Eden, shown as a mountain from which the four rivers of Paradise flow (the Ganges, Nile, Tigris, and Euphrates). As stated earlier, Columbus believed he had reached the mouths of these sacred rivers at the outskirts of Paradise when entering the Gulf of Paria, on the shoulder of South America, during his third voyage.[21]

Peering out from within the walls of Eden are two clothed men, rather than the figures of Adam and Eve which one would expect on a map of this heritage. Their identification is an enigma. Each holds an olive (?) branch, symbol of peace, suggesting that the pair may represent mutual tolerance and love between supposed adversaries, Jew and Christian. If this interpretation is accurate, the map's author was probably inspired by the thirteenth century Majorcan scholar Ramon Lull, who wrote of

> *"two men of marvellous wisdom, a Jew and a Christian,* [who] *esteemed each other in glowing love* [and, from a grove,] *for a long time they disputed in friendly conversation the advent of the Messias."* [22]

Both have bare heads, however, discouraging (though not eliminating) that amiable theory.[23] Conversely, the two might represent the allegorical source of all knowledge and rivers, the Master and his Novice. Immediately to the right (south) of Paradise lies *tabrobana* (Sri

---

21. See pages 47-48 in entry 12 (Ruysch).

22. Taken from Winter, "Notes on the Worldmap in 'Rudimentum Novitorum'," Imago Mundi, IX.

23. In the 1475 version both have bare heads and it is not clear if either wears a religious habit. This 1488 version more convincingly shows both in garments of a religious order, and the right-hand figure arguably (but not clearly) has a covered head. A 1491 rendering is altogether vague in these respects. The issue is confusing both because of the crudeness of the woodblock, and because of the occasional license of the colorist (which variable may have swayed Winter). Tony Campbell (The Earliest Printed Maps, p. 144), states that the figures are all bear-headed, none in the 1475 block wear religious habits, and only one in the 1488 block does so. See also Shirley, Mapping of the World, entry 2.

Lanka), the southeastern "corner" of the earth, which would soon be expanded to accommodate the discoveries of Columbus.

At the western end of the earth (bottom of the map) lie the Pillars of Hercules (Strait of Gibraltar), entrance to the western ocean. The southernmost point (i.e., 3 o'clock position) is *Ophir,* the Biblical place from where Solomon's ships brought gold and other riches. Just east of Ophir is the Phoenix, a bird believed to live in Arabia which regenerated itself after its five-hundred year lifespan by setting itself aflame on a pyre.

In the northeast (about 10 o'clock position) is a man who has just lost one arm to the devil. Travellers' stories warning of devils circulated in several variations in medieval lore. Most popular were the writings, circa 1360, of Sir John Mandeville (as he was known; he was doubtfully a knight, did not travel beyond Western Europe, and may not have been English).[24] Mandeville described a "valley perilous" in Mistorak (Armenia?) where the traveller, encountering tempests, thunder, and the sound of drums, entered a realm of devils. The story appears to have been plagiarized from Friar Odoric's account of his journey to China earlier in the fourteenth century, in which similar impressions were described in less sensational terms. Odoric's account was probably an honest but uncomprehending portrayal of natural phenomena: modern travellers have noted the huge (and devil-like) imposing rock figures in the highlands of Central Asia, and the apparent sound of drums which the acoustic nature of its topography seems to create.[25]

Directly to the west (below) lies a "sea of the amazons" *(mare amasoneorum),* an embodiment of an old tradition of a land in which only women dwell. Columbus would soon transplant the myth to the New World, and it would be recorded on such maps as that of Ruysch (entry 12).

The influence of the Crusades is found in the placement of the Holy Land at the center of the map, a common feature of post-Crusades *mappaemundi,* and in the figure of a king, holding a book, to the northeast of Ophir. This man is Prester John, whose mythical Christian stronghold was a brilliant hoax which became the focus of a search which helped motivate Renaissance exploration.

The myth of Prester John was fabricated by the fusing of two truths. The first was that there *were* in fact scattered pockets of Christians living in the East, principally the Nestorian Christians, whose ancestors in the mid-fifth century had fled the persecution they faced in Europe for their "unorthodox" beliefs regarding the enigma of Jesus being both man and god. Secondly, there also *were* formidable military powers in Asia who fought successful battles against Islam, most famously the 1141 assault against Persia by a Karezym /Kara Khitai coalition.[26] Following the First Crusade these circumstances were transformed into the myth of Prester John, an invincible Christian ally in the lands of the "infidels." Word of his kingdom was introduced into Western Europe after the First Crusade, apparently to allay its fears of engaging in further military intervention in the Levant. The lure of the tale was plain: no Christian army would be vulnerable to Islamic forces once it reached the impenetrable arm of the Kingdom.

That Second Crusade proved disastrous for Europe, however, and faith in the esteemed prester's existence might have diminished were it not for a new and more elaborate letter

---

24. Some scholars insist that "Sir John Mandeville" was a pseudonym for either Jean de Bourgogne or Jean d'Outremeuse, both of Liége. Others argue that the name was genuine, that a novelist named John Mandeville was born in St. Albans in the late thirteenth century and passed much of his life on the Continent, and that his book was only intended as a travel romance rather than as a factual account. The book, perhaps completed by 1356, was in any event commonly construed as a legitimate travel log.

25. See Malcolm Letts, Sir John Mandeville: The Man and his Book, Chapter IX.

26. Sanjar ibu Malik Shah of Persia had become threatening to Karezym, a neighboring kingdom; the Shah of Karezym approached Ye-lu Ta-Shih, prince of the Kara Khitai, to join forces, and together they overwhelmed Sanjar's forces near Samarkand. The Kara Khitai were of course not Christian; they probably followed some form of Buddhism. See also footnote 191.

circulated among the leaders of Western Europe, which this time —— most assuredly —— was written by Prester John himself. It was clear to the author of this *mappamundi* that Prester John's kingdom was located in Asia, for in that letter Prester John stated that

> *"our magnificence dominates the Three Indias, and extends to Farther India, where the body of St. Thomas the Apostle rests. It reaches through the desert toward the place of the rising of the sun, and continues through the valley of deserted Babylon close by the Tower of Babel."* [27]

Repeated European adventures in the East failed to locate the Kingdom, however, and as a result, as early as the 1339 chart of the Majorcan maker Dulcert, Africa became the preferred site. But as late as 1507 Waldseemüller still envisioned Prester John in the East, stating in the *Cosmographiae Introductio* (the same book that coined the name *America*) that Prester John is the one *"who rules both eastern and southern India and who resides in Biberith . . ."* [28]

That such an authority as Waldseemüller wrote with such credulity of Prester John well into the era of great discovery underscores the fact that we must not see this rude figure of a seated king as simply a medieval embellishment. Clearly, to the beholder of this map five hundred years ago, that figure represented a very real place on earth, a real but exotic world which transcended the mundanity of his fifteenth century Europe; according to Prester John himself, in his Kingdom the palace in *"which our sublimity dwells"* has a roof of ebony, gables of gold, gates of sardonyx inlaid with the horn of a serpent, windows of crystals, columns of ivory, and a courtyard paved in onyx. Further,

> *"our bed is of sapphire, because of its virtue of chastity. We possess the most beautiful women, but they approach us only four times in the year and then solely for the procreation of sons, and when they have been sanctified by us, as Bathsheba was by David, each returns to her place."*

And as for clothing,

> *"there is a kind of worm there which in our tongue are called salamanders. These worms can only live in fire, and make a skin around them as the silkworm does. This skin is carefully spun by the ladies of our palace, and from it we have cloth for our common use. When we wish to wash the garments made of this cloth, we put them into fire, and they come forth fresh and clean. Every kind of beast that is under heaven"*

can be found there, milk and honey abound, and *"no poison can do harm and no noisy frog croaks"* in this kingdom.

Just east of Prester John is the Tree of the Sun and Moon.[29] This oracular tree was shown to Alexander the Great while far into his conquest of the East. At dusk, the Sun Tree (which was masculine) and the Moon Tree (which was feminine) spoke to Alexander in an Indian language. He ordered the townsfolk to translate the trees' words, but they refused, for the trees had foretold of Alexander's death. In another version of the story, the two trees also spoke Greek, and told Alexander that he would die in May in Babylon by the hand of one of his own people, but refused to alter fate by revealing the name of the traitor.

---

27. Prester John letter quoted from Peter Forbath, <u>The River Congo</u>, p. 25-27.

28. "praesbyterum iohanne (qui et orientali & meridionali Indiae pracest atque in Biberith sedem tenet)," <u>Cosmographiae Introductio</u>, verso of leaf Aiiij.

29. This is apparently a variation of the "Arbre Sec" ("Dry Tree"), which was believed to have died upon the crucifixion of Christ. It was held that someday a prince from the west would travel to the Holy Land and sing a Mass beneath the dried tree, causing a miracle to occur : the tree would turn green once again and bear fruit, and many Jews and other "infidels" would embrace Christianity.

In the extreme northwest (at about the 7 o'clock position) lies the country of *viland* (or *vinland* on the 1475 block).[30] Rather than a record of pre-Columbian Norse settlement in North America, however, our *viland* more likely represents "Finland." Although extant textual references to Vinland date back to the eleventh century, its appearance on maps is otherwise unknown until the late sixteenth century.[31] Finland does not otherwise appear on printed maps until the Ziegler map of Scandinavia (1532), but it is found on manuscript maps antedating the *Rudimentun Novitiorum* and *Mer des Hystoires* by more than three centuries. The world map of the Arabic geographer al-Idrīsī (middle of the twelfth century) shows Finland near Russia and describes it as a most cold region having snowfalls of long duration —— clearly *Finland* rather than our American *Vinland*.[32] The Hereford *mappamundi* of circa 1290, to whose non-Ptolemaic geographic tradition the present map belongs, shows a group of three Baltic countries lying north of a bear representing Russia, of which the middle one is Finland, spelled the same as in the present map *(viland)*. Our *Viland* is almost certainly the child of these medieval Baltic traditions, not of pre-Columbian American landfalls.

- - - - - - - - - - - - - -

3. THE WORLD [hemisphere]

[untitled].

Circular map of the world after Macrobius, in the book *De Le Homini Illustri* . . . of Pliny. Siena, 1506.

Medium : woodcut.
Size of original : 125 mm diameter.

PLATE : 2

Although associated with the writings of Ambrosius Theodosius Macrobius, a Roman geographer, astronomer, and philosopher who flourished circa 399-423 A.D., this secular concept of the earth is rooted in the pre-Christian era. It is derived from Crates of Mallos (circa 150 B.C.), a geographer who headed the library at Pergamum, a highly cultural city in the northwest of what is now Turkey. Pergamum's library probably surpassed all except that at Alexandria. At about the same time as Crates' tenure, parchment was developed at Pergamum; the term "parchment" itself is in fact a corruption of the city's name. A century after Crates, the library was given to Cleopatra as a gift from Anthony.

Macrobius expounds the cosmography of Crates of Mallos by way of a commentary on Cicero's *Dream of Scipio,* which was part of the *De Re Publica* and was to an extent copied from Plato's *Vision of Er* from the *Republic*. Given the context, it will not be surprising that Macrobius' geographic concepts are poetic, and even philosophical.

According to Macrobius two ocean belts girdle the earth, one equatorial, the other along opposing meridians, thus dividing the earth into four equal parts, each with one "island" of

---

30. Written with the "n" as a top-script, consistent throughout the map.

31. With the exception of the "Vinland" map, purported to be a manuscript of circa 1440, itself a copy of older manuscripts. The authenticity of that document, which surfaced from an unknown (Spanish?) provenance through an American dealer in 1957, is still unresolved; it was "proven" fraudulent in the 1960s because of tests indicating the presence of titanium dioxide in the ink, but the accuracy of those tests has more recently been discredited.

32. The name, as transcribed from Arabic letters, is "bilād fimar" ("districts of Finland"). See Spekke, The Baltic Sea in Ancient Maps.

PLATE 2   Aurelius Macrobius, Siena, 1506   (entry 3)

land.[33] His map shows one hemisphere only, composed of the known world (the quadrant containing Europe, Asia, and Africa), and a southern quadrant, the Antipodes.

Macrobius believed that all four quadrants were inhabited. He proposed a clever counter-argument to critics who charged that, were there people "upside-down" in the Antipodes, they would simply fall off the earth altogether; rain, he pointed out, would then likewise all flow to the "underside" and fall off, making it impossible for even the northern hemisphere's oceans to contain their water. Humorously, he went on to speculate that

> *"the uninformed among the Antipodeans think the same about us, that we cannot be where we are because we would fall into the sky."* [34]

Nonetheless, some Church treatises demonstrated the absurdity of Antipodeans on a spherical earth with illustrations of people standing "upside-down" on the "underside" of the earth.[35]

From his view of the nature of our earth Macrobius held convictions discouraging the imperialist quest for fame and glory. There were both practical and virtuous considerations for this. Such conquest is a fickle goal, he explained, for no one could hold onto one's exploits long,

> *"owing to the floods and conflagrations that inevitably overwhelm the earth at regular intervals."*

Further, fame would be naive and deceitful since the earth itself is merely a minute and trivial point in the cosmos, and thus nothing of which we brag could really be important. Nor could one even flaunt one's self-proclaimed glory on the earth, except within one's own quadrant, for travel (and thereby communication) between quarters was prohibited by the oceans which separated them. People would forever live out their lives in their respective quadrants without any knowledge or contact with any other quadrant. Thus Macrobius' geography and cosmography gelled into a view of life: we must be content with what we have, our happiness would come only from within. Neither Columbus nor the five centuries of intensive exploration which followed him have been able to contradict Macrobius' philosophical ideals. Columbus, however, was successful in challenging parts of his cosmographical picture.

Macrobius believed that travel between either northern quadrant and either southern quadrant was impossible, not only because of the breadth of equatorial ocean which separated them, but additionally because that ocean was

> *"scorched with the heat of the sun."*

On that particular issue, Columbus, sailing south along African shores as far the Gold Coast in the early 1480's, had himself already proven Macrobius wrong.

But Columbus would ultimately fail to prove Macrobius wrong on two other points, both major issues which played against his arguments for a western voyage. First, Macrobius wisely adopted 252,000 stades as the circumference of the Earth, the figure which had been

---

33. Cicero described the earth as "a small island," which has sometimes been interpreted as implying that he believed it to be flat. Macrobius, however, states that Cicero had been misunderstood, that actually Cicero meant that "the entirety of the portion you inhabit" is a small island. Cicero's comment that our quadrant is "narrow at the top and broad at the sides" then makes sense as being one of four such "islands" on a spherical earth.

34. Macrobius, Commentary, book II ch V.

35. Notably Cosmas Indicopleustes (active circa A.D. 540). Cosmas was a merchant-trader in the Red Sea and Indian Ocean who converted to Christianity and devoted his life in a Sinai cloister to discrediting classical writings. He believed the earth to be a flat rectangle.

proposed in the third century B.C. by Eratosthenes.[36] This highly accurate figure was far too high for Columbus, who looked elsewhere for more attractive estimates.[37]

Second, Macrobius stated that east-west travel between either the two northern or two southern quadrants was impossible, owing to the expanse of sea between them (though —— unlike north-south travel —— not because this sea was "boiling"). Columbus, of course, disproved of this notion. But Macrobius' "other" northern quadrant correctly predicted that by crossing the western ocean Columbus would *not* reach the Orient, that another continent, a new world, lay in his way. Conceptually, Crates and Macrobius had invented an America, the "western antipodes."

The possibility that such undiscovered lands existed was openly considered by many intellectuals, particularly the Italian humanists, before Columbus' voyage, and had been the object of Portuguese expeditions to the west. The Church showed a mixed reaction to the concept of unknown lands, but overwhelmingly dismissed the possibility of unknown *inhabited* lands such as Macrobius described. It was clear to the Church, for example, that no inhabited land could exist that was unknown to the Apostles because Christ's teachings were supposed to have reached the entire world, and all surviving children of Adam were descendents of Noah's passengers and therefore all dispersed from a common point. These arguments were formidable for the devout Christian, and following Columbus' discovery of America they helped prolong the illusion that he had reached the Orient.[38] America, like Macrobius' Antipodes, violated these tenets.

The fact that Macrobius presented a circumnavigable Africa and an open Indian Ocean helped inspire Portuguese confidence in their own designs to sail south and east around Africa and ultimately on to the Orient. Macrobius also, however, unwittingly engineered a great tease. As can be seen on his map, Macrobius truncated Africa between the Tropic of Cancer and the equator, a latitude corresponding very roughly to the true southern coast of *West* Africa. As a result of this coincidence, Prince Henry's scouts, sailing generally due east for a naive thousand kilometers along those shores, "confirmed" Macrobius' optimistic picture of an easy sea route to India. In reality, of course, they still had more than half of the southerly breadth of the continent to sail, and the coast's bend to the south (present-day Cameroon) came as a bitter disappointment.

Some medieval *mappaemundi* do show what appears to be the Gulf of Guinea and southern Africa before its known exploration by the Portuguese.[39] This may have resulted from linking the southern coast of Macrobius' Africa to the northern coast of the Antipodes by means of a neck, either as a result of academic conjecture or the incorporation of outside (Arabic?) data.

In Roman times, controversy surrounded the nature of the Caspian Sea. Some authorities (e.g., Ptolemy, Aristotle, Herodotus) correctly considered it to be a closed body of water, but others (e.g., Pliny, Hecataeus) believed that it opened into the eastern ocean. Macrobius was convinced of the latter theory, and there was documented evidence to support it: the historian Pliny had recorded that two Romans, Seleucus and Antiochus, had entered the Caspian Sea from the Indian Ocean after sailing past India. Such is how it is shown on Macrobius' map, the large inlet of sea on the eastern end of Asia being the Caspian Sea.

This was one of several types of "zonal" *mappaemundi* which divided the globe into six climatic regions, or actually three different zones on both the northern and southern hemipheres. Straddling either side of the equator are hot zones (sometimes shown as a single

---

36. The modern equivalent of the Greek stade is uncertain; some authorities cite a figure of 148 - 158 meters, others of 185 meters, yielding a circumference of 37,296 - 39,816 km to 46,620 km. Eratosthenes' method was sound. Although the measurements he used were approximate, some of the errors appear to have cancelled each other out.

37. E.g., Ptolemy, Toscanelli, Behaim, d'Ailly.

38. And probably contributed to the viewpoint, common in the first half of the sixteenth century, that America and Asia were joined.

39. E.g., the Leardo map of 1448.

zone); midway north and south lie temperate zones; and at each polar region is a frigid zone. The inhabited land of each quadrant lies in the temperate zones.

Macrobius' text and map, frequently copied in manuscript during the Middle Ages, first appeared in printed form in 1483 in Brescia. The present 1506 rendering appeared in the writings of Caius Pliny ("the Elder"), the ancient historian mentioned above as having recorded a voyage into the Caspian Sea via the Indian Ocean. Pliny, who was responsible for many of the myths and legends which took root in medieval Europe, died of asphyxiation in A.D. 79 while investigating the famous eruption of Vesuvius which entombed Pompeii.

Place-names used in this rendition of Macrobius' map are: *idia* = India, *rubye* = Red Sea, *hieru* = Jerusalem, *alex* = Alexandria, *car* = Carthage. The book and map are also interesting as an early work to be produced in the city-state of Siena.

- - - - - - - - - - - - - -

Chapter III

CLASSICAL ENLIGHTENMENT:
THE RENAISSANCE BEFORE COLUMBUS

4. THE KNOWN WORLD

Generalis . . . [Strabo?] *descriptio.*

[anonymous], manuscript world map after Strabo, Italian?, circa 1550.

Medium : manuscript, pen and paper.
Size of original : 435 x 305 mm.

PLATE : 3

Strabo (circa 50 B.C.-25 A.D.), though himself a geographer of limited merit, compiled the views of many others who lived both during and before his time. He is particularly important for preserving the ideas of Eratosthenes (circa 276-196 B.C.). Strabo benefitted from Alexandria's pivotal role in long-distance commerce, which made it an excellent clearing-house for reports about far away lands. His view of the earth was a synthesis of both these reports and of the many learned writings he studied in Alexandria. He accepted Eratosthenes' estimate of the length of sea separating Iberia from Asia, and, contrary to Macrobius, stated that a mariner could theoretically make the crossing from one shore to the other.

The extent of land covered by Strabo is roughly the same as the *oikoumene* or "known world" quadrant of Macrobius.[40] The easterly breadth of Asia and the southerly breadth of Africa are severely truncated, thus with a finite coastline drawn for the entire world either controlled or sought by the Roman Empire of his day.

Strabo paints an encouraging prospect for anyone seeking a sea route to the East. Through the Strait of Gibraltar, named *Fretu Gadisanu* after the Phoenician/Roman name for Cadiz, a relatively manageable voyage would await a Mediterranean sailor around Africa, on to the Red Sea and thence to Sri Lanka *(Taprobane),* which was the southeast extreme of Strabo's *oikoumene.* The cape above Sri Lanka (marked *Colis*) is Cape Comorin, the southern tip of India, here the southeast corner of the continent. Traditionally, Strabo's Indus and Ganges Rivers are fitted into this geography by orienting the latter east-west, flowing into the Ocean Sea through Asia's east coast; however, the author of this manuscript has attempted to correct Strabo's Ganges by placing it on the southern coast (thereby allowing a more correct north-south orientation), but by doing so places it on the wrong side of India. On the eastern shore is *Seres,* the center of the silk-producing region of China and Tibet.

This late rendering of Strabo's concept of the world formed part of a mid-sixteenth century manuscript treatise. Its author traced cosmographic evolution from primitive zonal maps and a Ptolemaic cosmos through a competent current world map and a Copernican universe.[41]

- - - - - - - - - - - - - -

---

40. Hence the word "ecumenical."
41. See entry 28 and footnote 230.

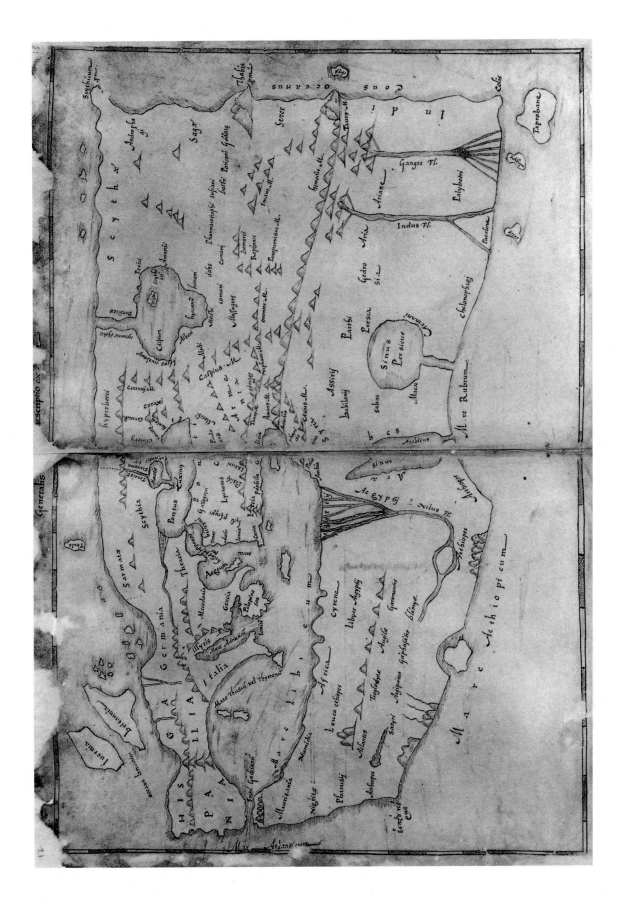

PLATE 3

Strabo,
Italian (?)
circa 1550

(entry 4)

5. THE KNOWN WORLD

[untitled].

Conrad Sweynheym and Arnold Buckinck, Rome, 1478.
Ptolemaic map of the world from Ptolemy's *Geographia*.

Medium : copperplate engraving.
Size of original : 320 x 540 mm.

PLATE : 4

The name of Claudius Ptolemy (circa A.D. 90 - 168) has come to symbolize the high state of geographic learning achieved by ancient Greek scholars. His *Geographia* was largely a critical adaption of the work of his contemporary, Marinus of Tyre (circa 100 A.D.). Although it was unknown to medieval Europe, it survived the Dark Ages and manuscript copies in Greek (without maps) circulated among the European intelligentsia by the early fifteenth century.[42] Its translation into Latin by 1406 and appearance in print by 1475 aroused Renaissance inquisitiveness. It was the most advanced cosmological overview found in a printed book at this time.[43]

In the *Geographia* Ptolemy described two projections for mapping the known world.[44] This map is constructed on his conical (first) projection, in which the meridians are straight lines which converge at the pole, and parallels are curved. It is typical of Byzantine models dating back to at least the thirteenth century. The second projection, which was more difficult to construct, had not been mastered until about ten years before this map and did not appear in printed form until four years after it.[45]

Ptolemy's world was much more sophisticated than that of Strabo, and the breadth of his *oikoumene* far more comprehensive. While Strabo had given Asia an arbitrary eastern shore by extending the east coast of the Indian subcontinent due north, Ptolemy maps the Indian Ocean fully through Southeast Asia, not attempting to chart Asia's eastern shores at all. The Indus and Ganges Rivers have assumed their correct positions, although the relative proportions of Sri Lanka and the Indian subcontinent are reversed. In place of Strabo's equatorial ocean Ptolemy has extended Africa through 16° south latitude and connected it with Southeast Asia by means of a "landbridge," rendering the Indian Ocean completely landlocked.

Wildly different theories have evolved to explain Ptolemy's geography of the eastern Indian Ocean. The easternmost peninsula in the Indian Ocean is *Aurea Chersonesus* ("Golden Peninsula"), named for the gold and alluvial ore reported to lie in its mountains and coast. This is the Malay Peninsula, or the Malay Peninsula and Indochina, with the Gulf of Siam being either the smaller bay in the southeast of the peninsula or the larger *Magnus Sinus* to the east. *Sinus Gangeticus* (i.e., Gulf of the Ganges) is the Bay of Bengal and Andaman Sea. *Magnus Sinus* ("Great Gulf"), if it is not the Gulf of Siam, would be the Gulf of Tonkin and

---

42. Thirteenth-century Byzantine manuscripts survive of Ptolemy's "Geographia," with the full set of world and twenty-six regional maps.

43. Knowledge of Greek was rare in western Europe. The project of translating the "Geographia" into Latin was undertaken by a Byzantine Greek scholar who settled in Italy, Emanuel Chrysoloras, and finished by a student of his, Jacobus Angelus, by 1406. The first known printed edition of Ptolemy's "Geographia" was published in Vicenza (northern Italy) in 1475, without maps. The first printed editions with maps are those of 1477 (Bologna) and the present Rome issue of 1478.

44. In the Geographia Ptolemy proposed a third projection for an armillary sphere. This projection was clever in that it allowed the known inhabited world to be seen unobstructed by the armillary rings. See Harley/Woodward et al., The History of Cartography, vol. I, p. 188-189.

45. See entry 6 ("Ulm" Ptolemy).

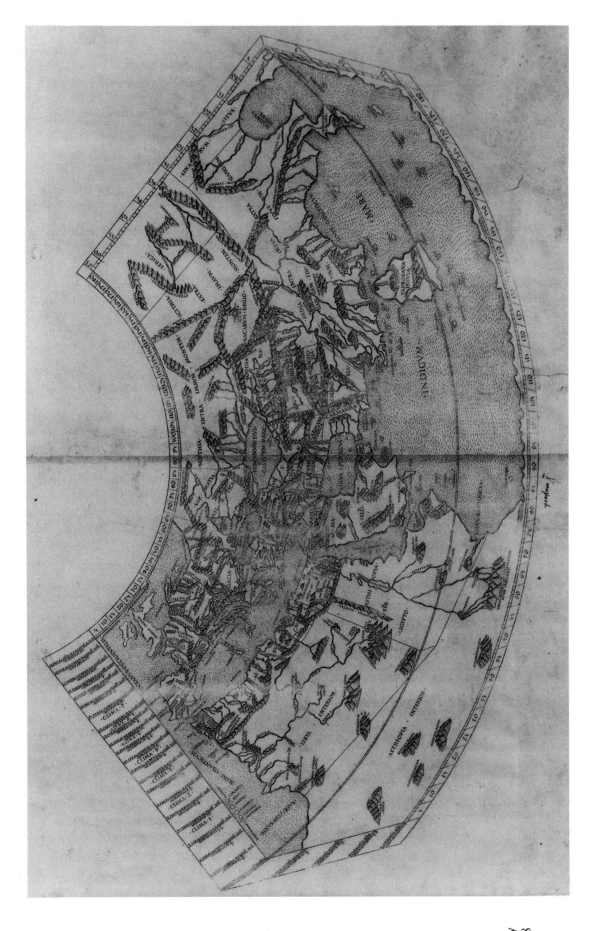

PLATE 4

Claudius Ptolemy
Rome, 1478

(entry 5)

the South China Sea. The larger question, though, is the meaning of the *Terra Incognita* landbridge extending from the eastern shores of *Magnus Sinus* to Africa. Most simply, it could be merely an error, occasioned by the sighting of the north coasts of Sumatra and possibly Java.

Another theory holds that Ptolemy deliberately landlocked the Indian Ocean as a ploy intended to help maintain Alexandria's monopoly on trade between East and West by dismissing any thought of an alternate sea route to India. Given Alexandria's stakes in the control of trans-Levant commerce (having laboriously constructed canals to enjoy what had formerly been a Phoenician monopoly), this is not far-fetched. When this map was printed in Rome one and a half millennia after Ptolemy, the Portuguese were pushing ever further along the African coast on the hunch that Ptolemy's landbridge did not exist.

Far more radical is the theory that Ptolemy's *Sinus Magnus* is the Pacific Ocean, and that its eastern shores is the west coast of South America. The first supposition, that Ptolemy's "Great Gulf" is the Pacific, is not new; Ortelius, for example, states on the verso of his *Maris Pacifici* that

> " [the Pacific] *was it not unnamed by ancient writers . . . Ptolemey falsly termes it Sinum Magnum, A great bay; whereas he should have nam'd it Mare Magnvm, A great sea.* "[46]

Ptolemy's designation of *Sinus Magnus* would thus make perfect sense, as would the indefinite southerly extension of its eastern coastline (that coastline bends to the left on the map because of the conical projection, but represents a continuation generally due south). This theory becomes even more intriguing when applied to Renaissance maps which open the Indian Ocean. Initially, the landbridge was severed from Africa but not deleted, left dangling as a mammoth Asian peninsula.[47] This peninsular vestige of the Ptolemaic landbridge helped reinforce a view that Columbus, upon reaching Central America on his fourth voyage (1502-03), had reached an Asiatic peninsula. Further, as early as 1448 a map bore an inscription on this land describing "giants who fight against dragons," a reference which has been correlated, if dubiously, to the "giant" Patagonian people later described by Pigafetta (chronicler of Magellan's voyage) as inhabiting southern South America.[48] Ultimately, the peninsular descendant of Ptolemy's landbridge did in fact assume the identity of a "true" South America, and a place-name from Ptolemy's landbridge even reached the Peruvian coast of a fully autonomous American continent.[49] But this transition does not in itself lend credibility to the theory that our *Terra Incognita* is South America and *Sinus Magnus* the Pacific Ocean.

Skipping east over the 180° breadth not shown by Ptolemy, the first place mapped is *Fortunatae Insulae,* the Canary Islands, the most westerly part of the world known to

---

46. Quoted from the 1606 <u>Theatrum</u> verso text of Ortelius' "Maris Pacifici" (entry 35). For correlation of the Africa-Asia landbridge with America see Dick Edgar Ibarra Grasso, <u>La representación de América en Mapas Romanos de Tiempos de Cristo</u>, Buenos Aires, 1970).

47. E.g., see items 11 (Waldseemüller), 13 (Apianus), 14 (Sylvanus), and 21 (anonymous /Holbein).

48. This inscription is found on the Walsperger map of 1448; see Paul Gallez, "Walsperger and his knowledge of the Patagonian Giants, 1448," <u>Imago Mundi</u>, Volume 33, p. 91-93.

49. Post-Ptolemaic Indochina functioning as America is found on the Finaeus maps of 1531 and 1534 (entries 19 & 20). The place-name from Ptolemy's landbridge which later found itself transposed to Peru is "Cattigara"; it is not shown in this 1478 rendering but is included in the 1482 Ulm issue (entry 6). The best example of its incarnation on American shores is the Münster map of America, 1540 (entry 23). Finaeus also used the term on American shores (entries 19 & 20), but the Münster is more compelling because he transferred the term even though he clearly separated the two continents.

Ptolemy and hence the point he prescribed for the prime meridian.[50] The Canaries had become the object of European colonization by the mid-fourteenth century, and later both the location and the prevailing winds of this extensive archipelago proved ideal for the west-bound Atlantic voyages. Columbus began all four Atlantic crossings from them.[51]

Two of Ptolemy's errors supported Columbus' proposals to the Spanish Crown. Ptolemy overestimated the breadth of the known world and undervalued a degree of longitude, and thus understated the distance between the west coast of Europe and the east coast of Asia, Columbus' most troublesome bit of unknown datum. But the anticipated distance had to be narrowed still further, and for that he turned to other sources.

Columbus owned a copy of this map in the 1478 issue of the Rome edition of Ptolemy's *Geographia,* begun by Conrad Sweynheym and printed, after the latter's death, by Arnold Buckinck. Sweynheym had been one of two Germans responsible for the founding of the first printing press in Italy, set up about 1464 in Subiaco (near Rome). In the latter part of 1467 Sweynheym and his partner moved to Rome where they pursued their craft, printing classical literature in quantity. By 1472 the new, narrow, and still elite market for printed works had been glutted, and they found themselves, in the words of their benefactor Bishop Bussi of Aleria in a plea for papal support, *"starving in a palaceful of unsaleable print."* [52]

In 1473, following a year of relief from the Pope, Sweynheym left the partnership. He is believed to have undertaken the atlas project at this time, or shortly afterwards. This Rome atlas pioneered the new medium of copperplate book engraving, and with it the problems of intaglio printing were resolved. Although a Bologna press published a similar work in 1477 (the year prior to the Rome atlas' completion), it is surmised that the Bologna printers hastily imitated the Rome work's technical solutions while it was still in progress. The elegance and finesse of the painstakingly engraved Rome atlas, though representing the medium in its infancy, was seldom rivalled. It epitomized the clean, spartan style for which Italian maps would become revered, the same aesthetic ideal already established by the "Italian style" of portolan chart making.

The atlas was reprinted in 1490 and 1507-08 from the same copperplates, with some difference in paper and watermark, and with subtle plate modification. On incunable strikes of this world map, the upper of the two lines representing the equator breaks for the island of *Bonae Fortvnae* in the Indian Ocean; by 1507-08 (as the present example) that upper equatorial line had been continued through *Bonae Fortvnae.*[53]

- - - - - - - - - - - - - -

50. In his <u>Geographia</u> Ptolemy used the most western point of the Canaries as the prime. In his <u>Almagest</u>, however, he recommended Alexandria as the prime, possibly a concession to diplomacy rather than science.

51. See footnote 6.

52. See Skelton's introduction to Theatrum Orbis Terrarum facsimile of the 1478 atlas.

53. Three factors make it difficult to state categorically when such a modification was performed: first, many of the surviving copies of the atlas are likely sophisticated (and so a bookseller, even a century ago, may have completed a 1478 atlas with a 1490 or 1507-08 map, or any combination thereof); secondly, even unsophisticated examples of the atlas may have been originally bound with an earlier loose strike of the map (and hence a "surplus" 1478 strike could have been originally bound with a 1490 atlas); thirdly, the plate may have been altered during, rather than between, editions.

6. THE KNOWN WORLD

[untitled].

Ptolemaic map of the world by Johann Schnitzer, published by Leinhart Holle. Ulm, 1482.

Medium : woodcut, with original hand color.
Size of original : 400 x 550 mm.

COLOR PLATE : II

This striking rendition of Ptolemy closely follows a manuscript of circa 1468 by Donnus Nicolaus Germanus, a Benedictine humanist.[54] Germanus is credited with two important contributions to the Ptolemaic picture, both of which are readily evident on this map when compared with the Rome work of four years earlier (entry 5).

The more apparent difference is the new projection. Having experimented with various refinements to the projections defined by Ptolemy, Germanus successfully tackled the spherical (second) projection which Ptolemy had described as superior but difficult to construct.

The other difference is geographic. Germanus supplemented the Ptolemaic *oikoumene* with better knowledge of Scandinavia after the Scandinavian geographer Claudius Clavus. Clavus' influence had been apparent on some Catalan and Italian charts since the first quarter of the fifteenth century, following a visit he made to Italy in 1424.

Some additional Ptolemaic nomenclature has also been added, of which *Cattigara,* on the Africa-Asia landbridge, is of interest for its later assimilation as a region of Peru.[55] Assuming the northern section of the landbridge to be Indochina, Cattigara would lie near what is now Hanoi. The Tropic of Capricorn was mistakenly labelled as the Tropic of Cancer in the woodblock, but has here been corrected in manuscript.[56]

This, the first edition of Ptolemy to be printed north of the Alps, has abandoned the medium of copperplate engraving pioneered in Italy in favor of the cruder woodcut. The woodcutter, Johann Schnitzer, signed his name on the woodblock (top center margin), thus introducing the practice of crediting a map's engraver.

The atlas was reprinted in 1486, with most of its maps unaltered except for the style in which they were colored. While blue is dominant in this earlier edition, the 1486 edition favored more golden-browns. This is the first instance in which an atlas was colored "in-house," to be offered for sale already illuminated by the colorist's brush. Uncolored examples are rare. Not until the *Theatrum* of Ortelius, nearly a century later, did the practice of in-house coloring become commonplace.

- - - - - - - - - - - - - -

54. Donnus Nicholas Germanus edited the Latin translation of the "Geographia" completed by Jacobus Angelus. See also footnote 43.

55. See entry 23 (Münster) and footnote 49.

56. This sort of oversight was not unusual in medieval maps. The magnificent "Hereford" world map, for example, has the names "Africa" and "Europe" reversed, yet the work itself was clearly a lavish production of high standards.

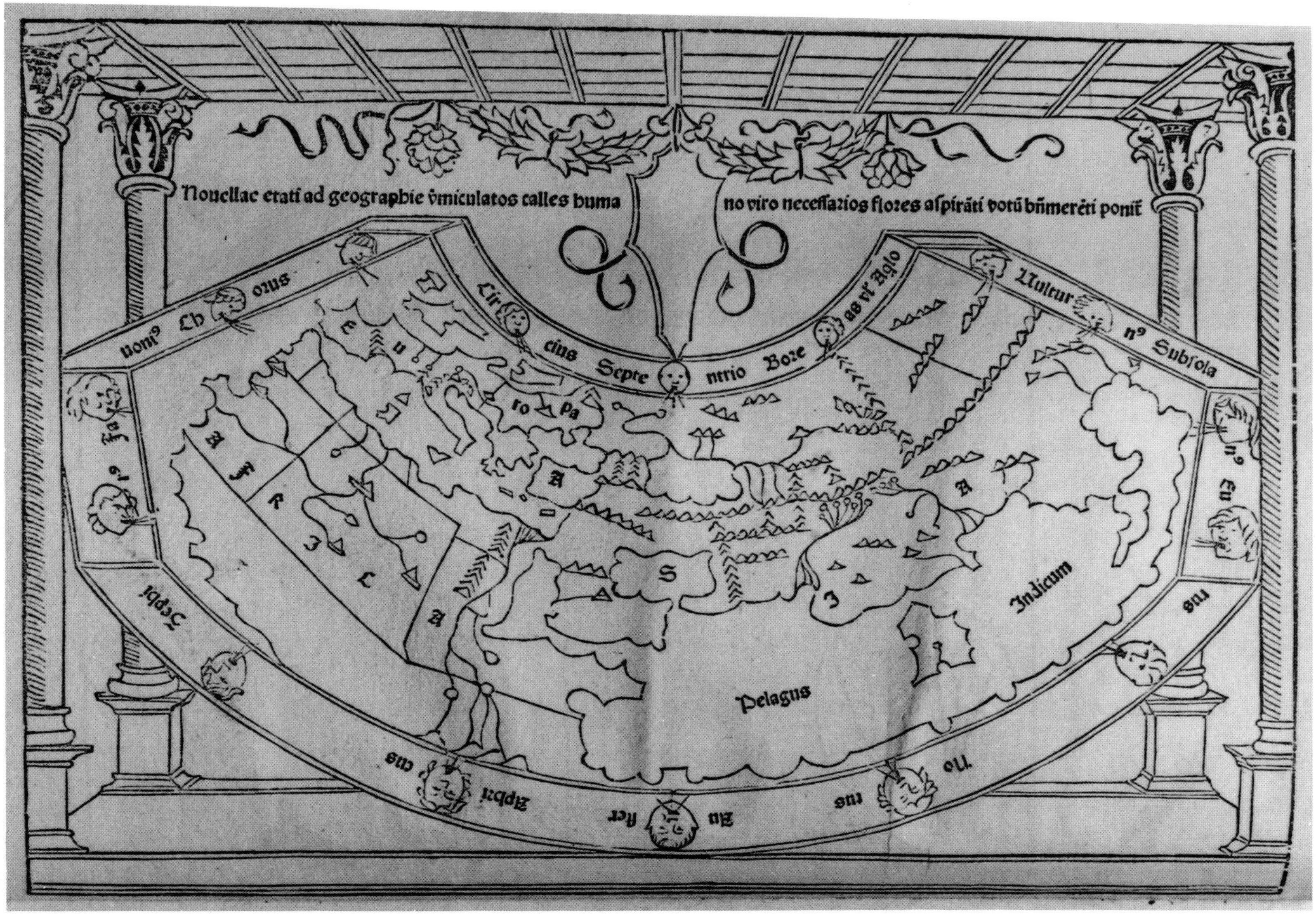

PLATE 5   Claudius Ptolemy /Pomponius Mela, Venice, 1482   (entry 7)

7. THE KNOWN WORLD

Novellae etati ad geographie umiculatos calles humano viro necessarios flores aspirati votubnmereti ponif.

Ptolemaic map of the world in Pomponius Mela's *Novellae etati ad geographie...* , Erhard Ratdolt, Venice, 1482.

Medium : woodcut.
Size of original : 135 x 185 mm.

PLATE : 5

Ptolemaic maps reached a wider and more varied audience by their inclusion in books other than Ptolemy's *Geographia*. The first such cases were this work of Pomponius Mela, and an atlas with text in poetry by Berlinghieri, both published the same year.

The simpler conical projection used in the 1478 Rome *Ptolemy* has here been combined with a "modern" Scandinavia similar to that found on the Ulm work of 1482. It is greatly simplified and encased within a border of Greek columns. The cosmographical treatise of Mela, a Roman geographer who lived during the very early Christian era (contemporary with Emperor Claudius, circa 40 A.D.), is a compilation of the views of Greek writers, hence the relevance of the map's unusual embellishment.

- - - - - - - - - - - - - -

8. THE KNOWN WORLD

Generale Ptholemei.

Modified Ptolemaic map of the world by Martin Waldseemüller.
Strassburg, 1513.

Medium : woodcut, with original hand color.
Size of original : 445 x 580 mm.

COLOR PLATE : III

Acknowledging Portuguese success in reaching the Indian Ocean via Africa, this Ptolemaic map of the world deletes the western segment of the landbridge which had classically connected sub-equatorial Africa to Southeast Asia. Although confined to the now-myopic breadth of Ptolemy's *oikoumene,* this map reveals what may have been in the minds of many explorers pushing westward over the New World in the early sixteenth century. The obsolete landbridge, left dangling from Southeast Asia, could now assume a new role as part of the New World.[57] In the same year this map was published, 1513, Vasco Nuñez de Balboa crossed the Central American isthmus and found the "south sea." Balboa may have believed he had actually crossed the mountains of *Cattigara* or its vicinity, the remaining section of the defunct landbridge. If so, when gazing out to the sea from the mountains of Darien he would have been sharing about the same panorama as Waldseemüller's wind-cherub *Favonius*. That Balboa perceived Darien to lie along this land is suggested by his term *South Sea,* which finally becomes logical in this context. From this viewpoint, Balboa was peering at an

---

57. The opening of the Indian Ocean had been pioneered by the maps of Henricus Martellus (circa 1490), Francesco Rosselli (1492-93), and Martin Behaim (globe, 1492). See also entry 20 (Finaeus).

austral sea. The oft-cited explanation that the isthmus at Darien is oriented east-west, and that Balboa was thus heading south rather than west when he reached the ocean, may have reinforced the reasoning rather than being its rationale.

To illustrate this point, compare the Indian Ocean of this work to the Balboan *Mare del Sur* on the Finaeus map of 1531 (entry 19), visible at about the 12 to 1 o'clock position on the left half of the double-cordiform. Balboa's "south sea" is part of the Ganges and Indian "Seas" (2 and 3 o'clock positions), which barely pierce the equator. As shown by Finaeus, only from the "other" ocean, *Mare athlanticum,* could one reach the Arctic region. Note that the meso-American isthmus near *Cattigara* from which Balboa sighted the Pacific is offset because of the projection, appearing at about 7 o'clock *(Dariena)*.[58]

Aside from its value in placing the newly modified Ptolemaic *oikoumene* in context of the evolving world image, Waldseemüller's opening of the old Ptolemaic landbridge is merely academic. The display of the most current knowledge was clearly not the intent of the map. This was the beginning of an anachronism that would survive yet another two centuries. Classical knowledge had become so highly venerated that the acceptance of new and better principles needed first to overcome its inertia. For example, notions and myths embedded in antiquity dampened acceptance of such monumental revelations as the Copernican universe and the circulation of the blood.[59] Similarly, if out of reverence rather than ignorance, Ptolemy's *Geographia* found a wide audience in the European public through the eighteenth century.

- - - - - - - - - - - - - -

58. In this volume, the author uses the term "meso-America" to denote the entire region of Central America and Mexico, through the northwest "corner" of the Gulf of Mexico (in present-day Texas). The modern definitions, in which North America technically extends south to the Isthmus of Tehuantepec, and Central America comprises the neck of land connecting Tehuantepec to South America, are inconvenient here. In days of earliest European exploration, all the land encountered on a westward run from the Spanish Main formed one component of the great puzzle; further, maps which present all of the New World as two Asian sub-continents (such as the maps of Finaeus) divide the land from the Gulf and south, Mexico thus forming part of the Central/South American sub-continent.

59. This dichotomy is found since the thirteenth century's Roger Bacon, champion of the scientific method yet defender of ancient knowledge. Copernicus' book dispelling the concept of a geo-centric universe was published in 1543; William Harvey's discovery of the circulation of the blood was published in 1628; the edition of Ptolemy's Geographia produced by Mercator in 1585, already a futile excercise at that date, was re-issued, with modifications and revisions to the plates but without geographic change, several times through 1730. That last edition is still common enough to presume that it enjoyed a wide appeal.

PART TWO — A NEW WORLD

The concept of a "new world" had incubated in the European subconscious long before the discovery of America. But precisely what did it mean for there to be lands heretofore unknown? There were two very different strains of thought on the matter.

The more acceptable foresaw the possibility, if not probability, of unknown islands or of previously unknown regions of the known continents. Various legends, some probably rooted in fact, told of islands to the west in the Ocean Sea. Once discovered, such islands were viewed either as extensions of the known world (such as Madeira or the Canaries) or as uninhabited, peripheral, isolated ones (the Azores). As for unknown shores of the known world, the precedent was most authoritatively set by Ptolemy. Ptolemy did not map the eastern coast of Asia, but it was there by implication, waiting to be found. It was this gap that early depictions of North America filled. These were new lands, new shores, but not a new world, and they could be integrated into the prevailing world image without radically disrupting it.

Conversely, a premonition of an entirely new world had also lain dormant in the Middle Ages. It appears principally in one of the most widely dispersed cosmographies in medieval Europe, that of Macrobius. Macrobius plainly stated that "antipodes" occupied both the northern and southern parts of the "other" side of the earth, in effect an entirely unknown universe co-existing with ours on the same earth. Following Columbus' landfall, various observers proposed that such a new world, a "western antipodes," had been found. Initially it was South America alone that became this Mundus Novus *(Columbus referred to it as "otro mundo") while North America was simply the unknown eastern end of Ptolemy's Asia, simply a new land* (Terra Nova).

Cosmographers and explorers shifted in and out of these possible realities, although the limitations of language make it difficult to know their thoughts precisely. Early on, Columbus declared that Cuba was the Aurea Chersonesus *of Ptolemy (Southeast Asia), but when Cuba's insularity could no longer be denied some cosmographers adopted the entirety of meso- and South America as that Asian peninsula. North America was the last to be recognized as a new world rather than a new land, being accorded continental proportions and tugged away from Asia in stages through the mid-sixteenth century.*

Chapter IV

THE WESTERN ANTIPODES?

9. PORTOLAN CHART

[untitled portolan chart of Europe, the Mediterranean and Black Seas, the coast of West Africa through the Canaries, and of undetermined islands in the Atlantic Ocean].

Attributed to Batista Beccari, Genoa(?), circa 1434.

Medium : illuminated manuscript on vellum, heightened in gold.
Size of original : 660 x 1170 mm.

COLOR PLATES : IV, V, and VI

The most accurate cartographic works dating from the early Renaissance were sea charts, or portolan charts, named for the "portolanos" (books of sailing directions, from "porto" [port]) with which they were used. Portolan charts had been used by pilots navigating the coasts of the Mediterranean and eastern Atlantic since at least the thirteenth century; the oldest surviving specimen, known as the "Carte Pisane," dates from circa 1296. Although the classic portolan chart depicted only the limited area known intimately by European sailors and did not confront the distant geographical or conceptual questions faced by maps of the world or hemisphere, its sophistication and detail are a marvel of human resource and ingenuity.

A typical portolan chart used by early Renaissance sailors delineated the Mediterranean and Black Seas more accurately than Ptolemy, and in fact better than found on most printed maps until the seventeenth century; such a chart correctly oriented Italy northwest-southeast (as opposed to Ptolemy's overly east-west alignment) and portrayed Mediterranean islands in better fashion; it presented a more correct trend to Scotland, and a superior depiction of Ireland.

Given its virtues, the portolan chart's failure to influence the *mappaemundi* of academia is baffling. It is possible that the portolan chart was considered the vulgar and uninspired tool of plebeian sailors, a base implement lacking in higher purpose, void of intellectual concept and silent on the matters of natural order. It may have been looked down upon as theologically barren, a prosaic, utilitarian labor which neglected the metaphysical. If this explanation is even partially correct, it would underscore that the concept of "map" as we now commonly understand it was still a minority definition in the early Renaissance. It would also suggest that in the portolan chart lie the conceptual beginnings of modern mapmaking.

The origin of the portolan chart is an enigma; it appears to have been born fully developed, being copied and recopied in various forms for three centuries. With only very specialized exceptions, portolan charts underwent little evolution. Of the theories advanced to explain this phenomenon, all are ultimately flawed or lacking in evidence.

A barely-viable hypothesis suggesting Neolithic origins represents the earliest, and perhaps most colorful, premise, followed chronologically by slightly less controversial claims of Phoenician or Egyptian origins. But the beginnings of academic acceptance lie in theories of classical origins. Passages in the writings of such ancient scholars as Strabo, Pliny, and Marinus of Tyre (via Ptolemy) have been construed as referring to sea charts. Finally, various medieval origins in Europe or the Near East have been advanced. Of these, some propose that portolan charts originated as simply an adaption of the information already contained in pilot books or portolans, or as a composite of many localized charts, or both.

Portolan charts are "skewed" because of magnetic declination, their authors taking magnetic north for true north. Since the pilot books with which seamen chose their course were uncorrected for declination as well, the error cancelled itself out and did not adversely affect navigation. Sailors had long been aware that a discrepancy existed between geographic and magnetic north, but neither understood its cause nor developed methods to correct it. The error itself may not have been significant when the compass came into general use in the Mediterranean in the thirteenth century.[60]

The present portolan chart is of northern Italian origin. Northern Italy had been a vital early hub in the art of chart-making, and, not coincidentally, much of Europe's early impetus to explore new oceans came from northern Italy. In particular, the cosmopolitan city-states of Venice, Genoa, and Florence had long been cultural, intellectual, and commercial pivots of medieval Europe and had accumulated much wealth through lucrative trade with the Levant and East. With their healthy economies and surplus capital, the Italian republics developed more sophisticated methods of money management. Such boons as letters of credit and double-entry bookkeeping allowed their resources to be channeled into a chosen enterprise, enterprises which by the end of the thirteenth century would include exploratory voyages.

The most celebrated merchant involved with Eastern trade was the Venetian Marco Polo, who spent two decades in the Orient between 1271 and 1295. The marvellous and influential account of his odyssey was committed to paper by a romance writer from Pisa named Rustichello, whom Polo met while the two were prisoners of Genoa during a war with Venice. As late as the 1560s, cosmographers turned to Polo's reports for clues to geographic riddles.[61]

After Polo, there came further (though far less perceptive or comprehensive) reports from other Italians. Among them, three who set out in the first half of the fourteenth century are notable. The first was Giovanni di Montecorvino, who in 1307 became the first Archbishop of Peking. Next, and most significant, was friar Odoric of Pordenone, who in 1316 embarked on a fourteen-year trek which took him first to India, then on by sea to China (circa 1320), where he spent three years in Peking, returning overland by way of Tibet and Persia; as a result of Odoric's reports, Canton (or *Zincalan*) can be found on some maps as early as the Catalan Atlas of circa 1375.[62] Lastly, there was Giovanni Marignolli, who was Papal Legate in Peking in 1342-46.

Various routes were followed by Italian traders, and commerce via the Levant continued despite the imposition by Moslem sultans of an Egyptian monopoly on trade through the Red Sea. Routes through central Asia and northeast Africa continued to provide Genoese merchants channels to the East, and routes through Mongol territory were safely travelled by such Venetian traders as the Polos. The need for a new and radical approach to Asia was, however, becoming increasingly apparent. This was first felt most acutely by the commerce-

---

60. The compass was definitely known in China at least by the twelfth century. In Europe, the mariner's compass may have been devised in Amalfi by the early twelfth century. Although that claim is disputed, a simple lodestone (magnetized needle suspended by a wood float) was definitely in use in Italy during the twelfth century. The degree of compass error existing at any particular time before 1500 has not been proven. According to Lloyd Brown (The Story of Maps, p. 132), it was slightly westward in the early thirteenth century, gradually reaching zero in 1350, then shifting to the east until 1655. It is also not clear whether there is a correlation between the magnetic declination existing at a particular time and place, and the degree of skewing on a contemporary chart.

61. E.g., the question of a strait separating North America from Asia; see entry 30 (Forlani /Zaltieri).

62. The so-called "Catalan Atlas" is a map of the world on twelve leaves mounted on folding boards. It was made by the Majorcan Jew Abraham Cresques, who was mapmaker to Pedro IV, King of Aragón. This is also the first map which was clearly influenced by the reports of the Polo's journey to the Far East. (Collection of the Bibliothèque Nationale, Paris).

based northern Italian city-states, and as a result it was they, not the Iberian powers, who made the first stabs at a sea route to the Orient.[63] Such an attempt is known to have been made even before the return of Marco Polo. In 1291, the same year Christiandom lost its only remaining footholds in the Holy Land with the fall of Acre and Tyre to Egypt, and the same year a fifty-three year (though largely ineffective) papal ban on trade with Egypt was instituted,[64] the Vivaldi brothers of Genoa sailed through the Strait of Gibraltar bound for "India." They followed the African coast at least as far as Cape Juby, in line with the Canaries, but were never heard from again and are presumed to have perished somewhere south of there. The Canaries themselves were reached by Lanzarotto, also of Genoa, about 1312.[65]

Due to Portugal's strategic location as the threshold of the western ocean, Genoese merchants and seamen became increasingly common in Portugal during the fourteenth century, particularly in the ports of Lisbon and Porto. Portuguese skill in the arts of navigation benefitted from the Genoese presence, and by the early fifteenth century they had supplanted the Italians as the pioneers in Atlantic exploration. Contact between the two nations remained strong, the reports brought by Portuguese explorers now cultivating the theoreticians of Italy. Beccari was partner to this Italian-Portuguese symbiosis during the reign of Portugal's master architect of exploratory strategy, the indefatigable Prince Henry the Navigator.[66]

Prodded and guided by Prince Henry, Portugal pursued two fronts, the circumnavigation of Africa, and the exploration of the western ocean. Tapping knowledge gained through both the earlier Genoese voyages and Henry's recent African ventures, Beccari charts the Canary Islands, and the coasts of what is now Morocco and the Spanish Sahara through Cape Bojador, just south of the Canaries. The fruits of Henry's western voyages are more perplexing, because it is not possible to reconstruct with any confidence just where his vessels reached. What we do know is that by 1427 Portuguese mariners, boldly sailing in search of islands to the west, had formally discovered the Azores. It is in such westward adventures as these that this portolan chart's riddles lie.[67]

---

63. Ancient attempts apparently were made to circumnavigate Africa, though without India as the objective. The famous instance is that recorded by Herodotus in which Phoenician navigators, under the Pharaoh Necho, sailed east-west around Africa circa 600 B.C. Herodotus dismissed this account because the sailors reported finding the sun on their right hand after setting a course north-west; Herodotus could not have known that such is precisely what would happen below the equator, effectively proving the account true.

64. Trade in war materials with the Moslems was already banned, although then, as today, even this taboo could be broken when convenient; Pope Nicholas IV, for example, allowed one Christian community to sell weapons to the sultan of Egypt because it was deemed necessary for its economic survival. See Scammell, The World Encompassed, p. 103, and Muldoon, Popes, Lawyers, and Infidels, p. 53-54.

65. Angelino Dulcert, in his portolan of 1339, records the island of Lanzarote in the Canary group and attributes its discovery to Lanzarotto Malocello (for whom it was named). Two years after Dulcert's chart, King Afonso IV of Portugal sent an expedition manned by Portuguese, Genoese, Spanish, and Florentine sailors south, reaching the Canaries again. The Canaries may have been intermittently "discovered" earlier.

66. The intercourse of ideas between Portugal and Italy continued after Henry's reign. The most famous instance was in 1474, when Portugal's African adventures had become frustrating and Afonso V turned to the Florentine scientist Paolo Toscanelli for advice (see introduction). That northern Italy continued to nurture exploratory talents is reflected in such people as Columbus (sailing for Spain), Cabot (for England), Vespucci (for Spain and Portugal), and Verrazano (for France), all of whom were from northern Italy.

67. Discovery of the Azores has been dated according to an inscription on the 1439 chart by Gabriel de Valseca; the date for discovery given on the inscription has been deciphered as 1432 or 1427 (the latter by Cortesão). Claims of fourteenth century Portuguese discovery of the Azores have, however, also been made. When colonization was undertaken circa 1445 (possibly under Cabral), Prince Henry first introduced domestic animals, then

Beccari records Atlantic islands in "blocks" with an outline encompassing an entire archipelago. He shows two pairs of blocks, each containing a large and a small group. This configuration is very similar to the earliest known chart to incorporate such islands, the 1424 chart of Zuane Pizzigano, except that Pizzigano orients them diagonally while Beccari shows them nearly straight north-south. Beccari has given names to each of the four blocks as a whole, rather than to individual islands; fading of the ink has made some difficult to discern. The least legible is the larger of the two northern groups, but by correlation to other surviving portolan charts (and specifically to another Batista Beccari chart) it would be *Satanagio,* the Genoese dialect equivalent to the Portuguese *Satanazes* (Satans).[68] To the north lies an umbrella-shaped block named *Tanmar,* and the island blocks to the south are called *Antillia* and *Royllo.* Only a few portolan charts survive from this early period showing these Atlantic islands. Their identity has been the source of controversy which will doubtfully ever be resolved.

The simplest possibility is that they are "false Azores," a cartographic fancy which helped inspire Henry's courage for western voyages.[69] This is unlikely, however, both because the portolan chart as a genre disdained speculation, and equally because the islands are accompanied by the statement *"Insulle de novo Rpto"* ("Islands newly discovered ['reperto']"). Some of the islands, specifically the northern ones, may represent "true" Azores. If so, this would verify Portuguese discovery of the Azores prior to the 1424 Pizzigano chart.

The two island blocks to the south, *Antillia* and *Royllo,* are particularly perplexing. Antilia is traditionally known as the mythical island of the Seven Cities, although that association was the result of the later merging of two separate traditions.[70] The term *Antilia* is probably a derivation of the Portuguese *ante* and *ilha* ("in front of" and "island"), or possibly of the Arabic *Al-tin,* the dragon.[71] It has also been flippantly associated with Atlantis, though with little support from academia.

There is early corroborative evidence of Portuguese landfalls in Antilia, whatever its identity; for example, Ruysch's world map of 1507 shows Antilia and states that it had been earlier found by the Portuguese. It was the opinion of the Portuguese historian Armando Cortesão that the Antilia of Henry's time represented fragmentary knowledge of the New World more than half a century before Columbus. Given the islands' latitude and position, and given Portugal's interest in the Canary Islands at that time, the possibility that they record Portuguese landfall in the Americas cannot be lightly dismissed. Portuguese presence in the Canaries was well established by the turn of the fifteenth century, with Spanish sovereignty over the islands not being mandated until 1479.[72] Though more obviously considered to be a base for exploration of the African coast, the winds off the Canaries are ideal for a trans-Atlantic crossing to the Caribbean, and a vessel venturing west of the islands by design or

---

Portuguese settlers; Flemish colonists arrived about mid-century. The islands may have been intermittently known in earlier times.

68. "ista ixolla dixemo" on the Pizzigano chart.

69. So-called "false Azores" are islands which may have been placed on early charts because of the widely held belief that such islands would be found in the western ocean. According to this theory, such hypothetical islands whet Henry's appetite for Atlantic exploration, rather than their being a record of actual discovery.

70. See page 43-44 in entry 12 (Ruysch).

71. Some early chart-makers certainly understood it as "ante ilha," as the island is found labelled as such ("Ante Yllas") on some portolan charts, e.g., in the "Miller" atlas of circa 1519. It interpretation as the Arabic "Al-tin" is found on the 1367 Pizigani map, which depicts an Atlantic island accompanied by an inscription and illustration about the Arabic legend of dragons. See A. Cortesão, "The North Atlantic nautical chart of 1424," Imago Mundi X.

72. The 1479 Treaty of Alcaçovas gave the Guinea coast, the Azores, the Madeiras, and Cape Verde Islands to Portugal, and gave the Canaries to Castille. Prior to that, the Portuguese dominated in the Canaries, and in fact had temporarily won official sanction for their claim with the bull "Romanus Pontifex" of Pope Nicholas V in 1454.

mishap could certainly have found itself washed up on Caribbean shores. In fact, the answer to our chart's Antilia may well lie in Columbus himself who, as stated earlier, used the Canaries as his point of departure for all four voyages to America. Were it possible to know whether he chose the Canaries by knowledge of earlier successes rather than by luck or political deference, it would then be possible to argue more convincingly that the Antilia and *Royllo* of Beccari and his colleagues were the "Indies" of Columbus. The prominence and relationship of the Canaries and Antilia/*Royllo* on Beccari's chart almost beg for this explanation.[73]

Another enigmatic land shown by Beccari is the island of *Brasil,* founded in Irish legend. Although the identification of Brasil is unknown, the consistency and coherency of surviving records supports the thesis of an Irish presence as far as the Faeroe Islands and Iceland by the early ninth century, with sketchy evidence suggesting landfalls further west.[74] Brasil island appeared on portolan charts as early as the Dalorto map of circa 1325, charted as an island to the west of Ireland (as it is shown here by Beccari), though it subsequently moved about in the Atlantic at the whims of Italian and Catalan chart-makers. Theories as to the origin of this island's name also remain guess-work. It might be derived from the Gaelic word for "fortunate" or "blessed," or from the Romance word for "brazier" (brassworker), or from its root "bras," meaning "flame-colored," which in this case would refer to dyewoods, though to species other than those for which the Brazil of South America was named.[75]

The Ruysch map of 1507 shows an island to the northwest of Ireland with an inscription in Latin stating that *"in 1456 this island was completely consumed by fire."* As the features accorded Brasil on some charts make it look volcanic in nature, the association with Ruysch's lost island has invited speculation. Icelandic tradition, in fact, records an island between Iceland and Greenland, which does not exist in our time, from which the coastal mountains of both were visible.[76] Brasil continued to appear on British Admiralty charts as "Brazil Rock" as late as 1873.

In a fashion typical of the more elaborate portolan charts, Beccari has adorned the otherwise void inland areas of his chart with vignettes of the cities considered most important by him, the most prominent being his native (?) Genoa. Others are: Venice, Genoa's perennial rival; Santiago de Compostella, in northwest Spain, the most important place of pilgrimage in medieval Europe after Jerusalem and Rome; Marseille, an important commercial center and a departure port for the Crusades; Cologne, commercial center and river port; Cairo, seat of Egyptian control over Levant trade; and ever-important Jerusalem. The remaining two are the least familiar to modern eyes. One is Varna, lying on the shores of the Black Sea. Varna had been captured by the Turks in 1391, and by Beccari's time had become an important Ottoman port. Within a decade of this chart (1444) Varna was the site of Europe's final offensive defeat in her battle against the Ottomans, with her loss of the Near East soon to be consumated by the fall of Constantinople in 1453. The remaining

---

73. However, attempts to correlate these islands to specific points in America, no matter how tediously researched, amount only to interesting speculation. For example, Kelley, in Imago Mundi 31, suggests that "Royllo" is Norman or Breton for a stockade, thereby indicating, in his view, a fortified Indian village. Likewise "antillium" was apparently an old term for "a shield," again suggesting that Antilia was a fortified villlage. According to Kelley "Satanazes" would be Nova Scotia and the Bay of Fundy, and "Antillia" the coast from northern New Jersey to North Carolina.

74. Irish knowledge of the Faeroe Islands and Iceland is recorded as early as A.D. 825 in Dicuil's De Mensura Orbis Terrae.

75. Brazilwood is not native to islands in the eastern Atlantic, and the possibility of early Irish contact with West Indian dye-wood sources is remote. Brazilwood from the East Indies was, however, an object of Asian trade in medieval times. If this derivation is correct, it may have referred to milder dyes made from lichens, known as "archil." For Brasil's interpretation as "fortunate/blessed" see e.g., Sauer, Northern Mists, p. 167-169 ; as "brassworker" see J.E. Kelly, Non-Mediterranean Influences that shaped the Atlantic in the early Portolan Charts, Imago Mundi 31.

76. See Nordenskiöld, Facsimile Atlas, p. 65.

metropolis illustrated is Fez *(Tirimissen,* offset to the east) which for Europe was perhaps the most fabulous and exotic of these cities depicted in vignette. Founded in 808 A.D., in Beccari's time the "new city" of Fez, connected by walls to the old, had been built, and Fez had already reached its greatest glory under the Marinid sultans of the fourteenth century. At the time of Beccari's chart, Fez had taken on particular significance for Europe because of its proximity to Ceuta, whose conquest by Portugal in 1415 marked the first permanent European foothold in Africa and the beginning of European overseas expansion. Below Fez, along the bottom of the chart, an inscription explains that the region to the south consists of deserts and great forests, and is inhabited by black people.

Little is known about the maker Batista Beccari. On this chart the vignette of Genoa is clearly more prominent than any other, suggesting that he worked in, or at least considered his allegiance to be to, that city-state. That Batista worked in Genoa is also strongly supported by a document which recently surfaced.[77] This document is a contract in which Batista, residing in Genoa, agreed to apprentice a nine-year-old boy named Raffaelino Sarzana in the art of making charts for a period of eight years. The document is dated August 17, 1427, placing the end of Raffaelino's apprenticeship term at 1435, just after the tentative date assigned to this portolan chart.

Another document, better known and equally compelling, suggests that Batista was *not* present in Genoa shortly *after* the pivotal year of 1435. In a petition dated November 7 of 1438 Agostino da Noli, a chartmaker for whom no extant work is known, requested a reprieve on taxes owing to his poor financial state. In his request, which is addressed to "Illustris et excellentissimus dominus Thomas de Campofregoso," da Noli described himself as being the only chartmaker currently working in Genoa.[78] In the Council of Elders' reply of December 2, da Noli was granted a ten-year remission of taxes with the stipulation that he was to teach the art of chart-making to his brother. That the state consented to da Noli's request, and that they made it contingent on his training another in the same field, both suggest that he was indeed the sole practicing chartmaker in Genoa at that time, and therefore that Batista Beccari was no longer resident in Genoa. Given Batista's obvious interest in new explorations, it is reasonable conjecture that sometime before 1438 Beccari entered the employ of Prince Henry, whose envoys at this time combed Europe to recruit its finest geographers to serve the school of navigation he is believed to have founded at Sagres.

A reference to Batista is found in an inscription on a portolan chart of 1447 by Piero Roselli, who was claimed both by Spain and Italy, worked in Majorca, and created charts in the Catalan style. Roselli states that he conceived the chart *"de arte Baptista Beccarii."* This phrase, beyond showing that Roselli held Batista in high regard, suggests that Batista had been Roselli's mentor.

Batista was probably the son or relative of Francesco Beccari, a chartmaker who was active at the turn of the fifteenth century and who is known to have been working in Barcelona in 1399. Both Francesco and Batista were pioneers, a new breed of chart-maker, anxious not only to improve the delineations of familiar coasts but also to extend their charts' frontiers. Both makers are highly regarded for the originality of their typonomy and their cartographic innovations. Batista, for example, appears to have introduced the practice of emphasizing coasts by color, and adopted an improved rendering of Atlantic distances and Sardinia introduced by Francesco.[79] As the new data concerning the Ocean Sea which the

---

77. See Tony Campbell's "Portolan Charts from the Late Thirteenth Century to 1500" in The History of Cartography, p. 431, n. 415.

78. See Revelli, Cristoforo Colombo, p. 461-462.

79. Batista's accentuating coasts by color was first noted on his portolan of 1426 (Bayerische Staatsbibliothek, Munich). The improved charting of Atlantic distances and of Sardinia were apparently introduced in a 1403 chart of the elder Beccari. In an inscription on that chart he explains that, owing to complaints and advice from seamen, he has lengthened Atlantic distances and adjusted the position of Sardinia. Francesco's address, as translated in H. P. Krauss Catalogue 95 (1961), states in part that: "Franciscus Becharius . . . in this and other charts . . . from after A.D. 1400, lengthened the distance of the coasting

present chart records was of little pertinence to traditional ports-of-call, its intended audience, rather than a typical sailor, must have been a land-bound theoretician, a government agent, or the pilot of an experimental voyage into the western ocean.

- - - - - - - - - - - - - -

10. THE KNOWN WORLD

[untitled].

Gregor Reisch, Freiburg, 1503. Ptolemaic map of the world from the *Margarita Philosophica.*

Medium : woodcut.
Size of original : 280 x 410 mm.

PLATE : 6

Here we find the first hint of knowledge of Columbus' discoveries on a printed map. Though Ptolemaic in its geographic content, this map contains a legend which alludes to the discovery of the New World. That legend appears on the Ptolemaic landbridge connecting Southeast Asia to Africa. It states, in Latin, that

> *"here there is not land but sea, in which there are such islands not known to Ptolemy."*

The islands to which it refers could be either those of Australasia, the fabled Spice Islands, or, more likely, those of the Caribbean, the Spanish Main, which at this time was equated with the South China Sea.

Twelve windheads surround the map and blow the various winds onto the earth. Reisch's windheads are quite different from the reserved cherubic faces which perform this task on earlier *mappaemundi*. In true Renaissance spirit these windheads are vibrant and charismatic, all individualistic and distinct from one another. One of particular note is *Vulternus,* gazing towards Asia with spectacles, this being among the earliest depictions of eye glasses.[80]

This transitional map appeared in some editions of the *Margarita Philosophica*, a widely read encyclopedia of moral and natural philosophy.

- - - - - - - - - - - - - -

navigation . . . especially in the coast of Portugal, viz, from Cape St. Vincent even to Cape Finisterre, and . . . Vizcaya and the coast of Bretagne and of the island of England . . . the marrow of the truth having been discovered concerning these (things) aforesaid through the efficacious experience and most sure reports of many, i.e. masters, ship-owners, skippers and pilots of the seas of Spain and those parts and also of many of those who are experienced in sea duty, who frequently over a long period of time sailed those regions and seas . . . (therefore) I placed the said island (of Sardinia) in the present chart in its proper place where it ought to be." The improvement had been noted on a Batista portolan of 1435 (dated 1436) in a report by Hermann Wagner published in 1895, well before the discovery of Francesco's earlier chart with its explanation.

80. Although the item itself can be traced back to Roger Bacon in the thirteenth century, and possibly earlier in some Asian cultures.

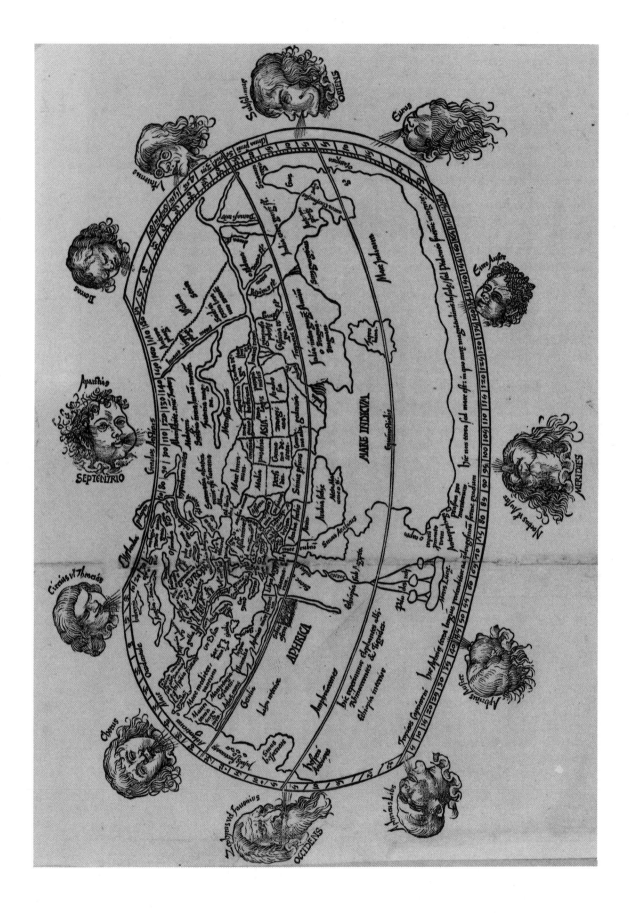

PLATE 6

Claudius Ptolemy/
Gregor Reisch
Freiburg, 1503

(entry 10)

11. THE WORLD

Orbis Typus Universalis IVXTA Hydrographorum Traditionem.

[A Map of the Entire World, According to the Teachings of Hydrographers]. Martin Waldseemüller, circa 1505-06 (?) (but Strassburg, 1513).

Medium : woodcut, with original hand color.
Size of original : 445 x 570 mm.

COLOR PLATE : VII

In 1498 Columbus, on his third voyage, sailed south of his earlier exploits and discovered a land he said was heretofore unknown. It was called *Paria* by its inhabitants. Encountering a river (the Orinoco) which discharged an enormous volume of water into the sea, the Admiral concluded that this land was continental. Although he persisted in the exuberant delusion that he had reached the outskirts of Paradise, for the first time Columbus' oft-repeated claim of having reached a continent was true. Remarkably, he also decided that the continent was a new world *("otro mundo").*

Two years later, in 1500, Manoel I of Portugal sent Pedro Cabral on an expedition to India via southern Africa, anxious to repeat the success of Vasco da Gama, who had returned from the first successful sea crossing to India the previous year. Cabral, deliberately or inadvertently, strayed to the west and discovered Brazil.[81] He sent one of his thirteen ships back to Lisbon to announce the find.

Mapmakers quickly realized that Columbus and Cabral had discovered different coasts of the same land, a new continent with soil on both the Spanish and Portuguese sides of the papal demarcation line.[82] It quickly made its way onto maps in the general configuration used here by Waldseemüller.

On Waldseemüller's map, the most northwesterly feature marked on South America is the Gulf of Venezuela *(baroia).* Above it are two islands named *giga*[nti?] and *brasil,* the latter a curious bridge between the island of Irish myth and "real" Brazil, the land of Cabral named for its dyewood. Continuing east, the next name is *Canibiles* (Cannibals) at the Gulf of Paria, Columbus' gateway to Paradise. The cape above modern-day Recife is *Captit Ste Cruns* (i.e., "Cape of the Holy Cross"), a name given by Cabral on April 23 of 1500. Furthest south is *alta pago de S. paulo* ("The village of St. Paul"). Above South America lie the islands of Cuba *(isabella),* Jamaica (unnamed), and Hispaniola *(Spagnola).*

The map's only trace of North America lies far to the northeast, where a landfall, probably from a Corte-Real voyage, is recorded.[83] In the title Waldseemüller associates his map with *hydrographorum,* hydrographers, i.e., a sea-chart, and in the style typical of its genre he has avoided conjecture. As a result, the western and southern shores of North and South America are left uncharted, their nature and extent still quite unknown. Further, Waldseemüller does not show a full 360° breadth of longitude, enabling him to clip virtually the entire Pacific Ocean from his map. He does, however, give Asia a finite eastern coast, and by doing so has asserted clearly that both North and South America, whatever their nature, are in fact new lands quite distinct from Asia.

His depiction of Asia itself displays an obvious familiarity with Portuguese advances into the Indian Ocean, as he breaks with Ptolemaic tradition by showing the Indian subcontinent

---

81. A 1497 voyage claimed for Vespucci might have reached Brazil, but the authenticity of that particular voyage is doubtful.

82. The 1494 Treaty of Tordesillas divided the non-Christian world between Spain and Portugal. It established the "raya," the line of demarcation, at the longitude 370 leagues west of the Cape Verde Islands. See also footnote 144.

83. See page 56 in entry 14 (Sylvanus) for Corte-Real.

and Sri Lanka in their true relative proportions. In Southeast Asia, east of the Malay Peninsula, there is an "extra" mammoth peninsula extending south to beyond the Tropic of Capricorn. This is a vestige of Ptolemy's Africa-Asia landbridge and was an attempt to reconcile that landbridge with the realization that the Indian Ocean was not the landlocked sea described by Ptolemy. Bartolomeo Dias' voyage of 1487-88 around the Cape of Good Hope had effectively laid that Ptolemaic error to rest.[84]

The question of this map's rightful niche in history rests precariously on the uncertain date of its creation. Although not known to have been published until its inclusion in Waldseemüller's atlas in 1513, evidence suggests that it was prepared at an earlier date. Four considerations favor this: *(1)* indications that work on the atlas had begun in 1505 or 1506, then being abandoned because of financial troubles; *(2)* the existence (in a single example) of this map with the name *America* inserted on the woodblock denoting the New World, apparently struck before Vespucci fell from Waldseemüller's grace shortly after 1507;[85] *(3)* the fact that the 1513 atlas' separate maps of America and Asia are geographically more advanced than this map's rendering of those continents;[86] and *(4)* the fact that this map is not uniformly sized with the other maps in the atlas, being larger and as a result often clipped in the binding. A strong, but inconclusive, argument against the most optimistic dating of 1505-06 is that the map's rendering of India and Sri Lanka is "modern," while they were still shown in Ptolemaic fashion on Waldseemüller's large map and gores of 1507.

If in fact the map was conceived prior to 1507, it would represent the earliest depiction of North America as separate from Asia, and if prior to 1506 it would be the earliest printed map to show the New World at all.[87] By 1513, however, this Waldseemüller map was a dated work, of interest only because of the scarcity of maps from the early post-Columbian period.

- - - - - - - - - - - - - -

84. First shown after discovery in the manuscript maps of Henricus Martellus, Nuremberg, circa 1490 (British Library and Yale University), although many pre- da Gama maps, free of Ptolemaic influence, hypothetically show an open Indian Ocean.

85. This example of Waldseemüller's map with the name "America" may pre-date his large map and gores famous for use of the term. It is in the collection of the John Carter Brown Library. See Henry N. Stevens, The First Delineation of the New World and The First Use of the Name America on a Printed Map, London, 1928. See entry 13 (Apianus) for synopsis of Vespucci and Waldseemüller.

86. E.g., the world map's repeated Southeast Asian peninsula was corrected on the atlas' "modern" map of Asia, and its separate map of America is much more comprehensive.

87. The earliest printed map of certain date showing any part of the New World is the Contarini-Rosselli world map of 1506, known by the example discovered in 1922 and now in the British Library.

12. THE WORLD

Universalior Cogniti Orbis Tabula. Ex recentibus confecta observationibus.

[A Map of the Known World, according to the newest discoveries]. Johann Ruysch, Rome, 1507. Added to the 1508 issue (and some 1507 examples) of the 1478 "Rome" *Geographia* of Ptolemy.

Medium : copperplate engraving, two sheets joined.
Size of original : 405 x 535 mm.

PLATES : 7 and 8

England's first window to the open Atlantic was the port of Bristol, in the southwest part of the country bordering Wales and facing Ireland. By the early fifteenth century, Bristol ships dominated north Atlantic commerce. Her mariners made the shuttle to Iceland to get fish, particularly herring, entered the Mediterranean to fetch wine from southern France, and added Madeira to the itinerary once Portuguese colonists began to produce wine and sugar.

By about 1480, however, a curious twist can be detected in Bristol trade. The legendary Irish island of Brasil, long rumored to lie far to the west of Ireland, began to occupy the fancies of some Bristol sailors, and at the same time huge quantities of salt are recorded as their west-bound cargo. As the principal maritime use for salt was the curing of cod, the inference is that Bristol fishermen had sailed beyond Iceland and located the bountiful codfish sources off the Newfoundland coast.[88]

In any event, Bristol soon became the spring-board for more ambitious projects. In 1497 John Cabot, of Venetian citizenship and Genoese birth, sailed west from Bristol in an attempt to reach China, making landfall in the northeast of North America.[89] The Netherlander Johann Ruysch, author of the present map, is believed to have participated in a Cabot or similar Bristol-based English expedition to America. His is the first published map made by an actual explorer of the New World.

Ruysch's observations in the New World may have influenced, although it is doubtful whether they were the basis for, his mapping of the tiny portion of northern American shores he visited. He depicts North America as an easterly protrusion of Asia, poetically, though inadvertently, appearing as a flower bud ripe to blossom, which, in 1507, indeed it was.[90]

Neither Ruysch's adventures nor those of his fellow Bristol sailors could have had anything, however, to do with any other part of the map. Even the northeast coast he himself visited shows the influence of the Portuguese, who by early in the century had explored eastern Canada and had come to dominate its fertile fishing grounds.

---

88. While the search for Brasil may have initially helped lure Bristol sailors further west, it ultimately may have served to camouflage their discovery of Newfoundland's immensely lucrative waters. Bristol records show, for example, that in 1481 a Bristol consortium obtained a duty-exemption for forty bushels of salt for each of two ships sent to find and examine "a certain island called the Isle of Brasil." Records verify that the ships returned, but nothing further (see Sauer, Northern Mists, p. 36). Cod was an ideal fish both because it was immensely popular from the culinary standpoint, and because the curing methods used to preserve it were far simpler than those needed for more oily fish such as herring and mackerel. Cod was usually preserved with salt, and could even be dried on board ship.

89. Opinions as to where Cabot reached range from Greenland to Newfoundland to Nova Scotia to New England. Hakluyt, a late sixteenth century promoter of British expansion and clearly a biased observer, thought Cabot sailed fully south to Florida.

90. This configuration is reminiscent of the Contarini-Rosselli map of 1506 and is ultimately rooted in the "Cantino" map of 1502 (manuscript; see also footnote 135).

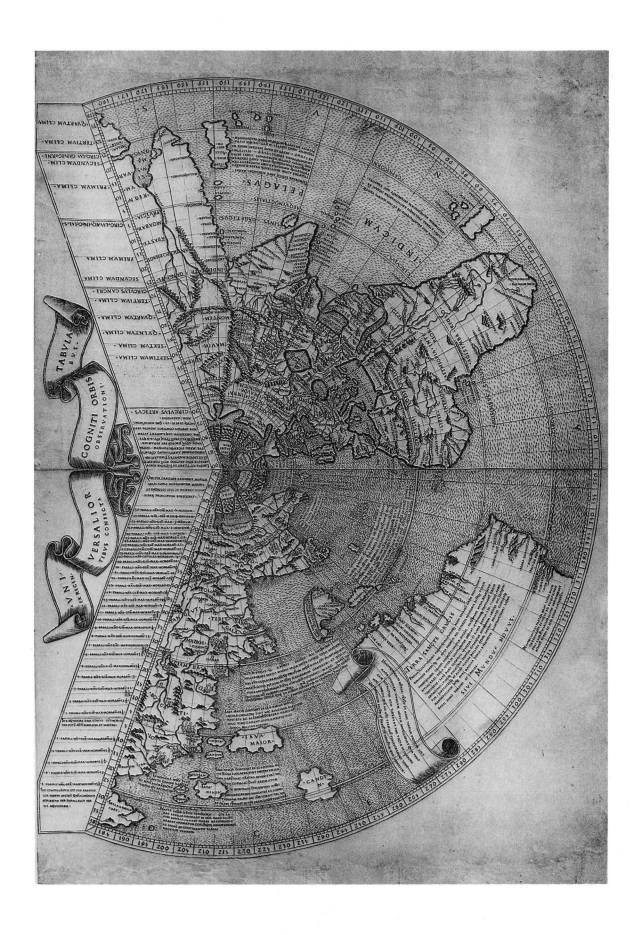

PLATE 7

Johann Ruysch
Rome, 1507

(entry 12)

In fact, by 1506 Portuguese fishermen were hauling so much cod from northern American waters that their monarch sought tariffs on its import.[91] In its demand for revenues, the Crown cited the land from which the fish came as *Terra Nova,* referring to the land explored by the two Azorian/Portuguese Corte Real brothers in 1501-02; the king of England used the same phrase ("New found land") for the land discovered by John Cabot. Ruysch uses the term to designate the entire peninsula of North America, and names its most southeasterly point as *C. DE PORTOGESI,* acknowledging Portuguese dominance there. Just north of that cape is *IN. BACCALAURAS* ("codfish island"), testifying to that fish's profound influence on early Atlantic exploration. This term, in its sundry Romance variations, soon became a major place-name associated with North America.[92] To the north, at about 48° north latitude, is *R GRADO.* The editor of the atlas, Beneventanus, clarifies this ambiguous term as *Caput Grande.* This comment, as well as the river's latitude and its probable origin in the Corte-Real voyages, all suggest that this is the Gulf of St. Lawrence. The adjacent features also make sense in this interpretation: *C GLACIATO* (a reference to glacial waters) would be Newfoundland, *C DE PORTOGESI* would be Nova Scotia, and *IN BACCALAURAS* would be Cape Breton or Prince Edward Island.

In the north, our theoretical Gulf of St. Lawrence is *BAIA DE ROCKAS* (possibly lower Strait of Belle Isle). The use of the letter "k" in this "Rocky Bay" is one of the map's scant traces of the British flag under which Ruysch is believed to have sailed. An inscription far to the north states that

> *"Here begins the Sugenum Sea* (i.e., "whirlpool" sea). *Here the compasses of the ships lose their power, and it is not possible for ships which have iron on board to return."* [93]

This is sometimes interpreted to be another element of British influence, as the variation of the compass orientation had been noted by Cabot on his second voyage. More likely, however, the legend originated in a medieval treatise, the same from which Ruysch configured his Arctic region.[94] Along the Asian coast facing this whirlpool arctic is *IUDEI INCLUSI,* where the lost tribes of Israel were thought to dwell. In the Atlantic between Ireland and Greenland lies an island with an inscription stating that

> *"this island in the year of our Lord 1456 was totally consumed by fire."*

It is tempting to suggest that this is an epitaph for the fabulous island of Brasil from Irish legend.[95]

Greenland *(GRVENLANT),* like North America, is depicted as an extension of the Orient. An inscription describing nearby islands warns that sailors who had gone to them had been tricked by demons. Inland from North America, into "Asia" proper, the influence of the medieval imagination is found in the dreaded realm of Gog and Magog from which the Antichrist would spring at Armageddon. Originally a Biblical saga, the story of these two

---

91. In 1506 the Crown sought control of the "tithing" of the cod trade (the levying of a 10% tax) which had already been effected by the officials of the northern province of Entre Douro e Minho. Although Portuguese precedent in the region is generally set with the voyages of Gaspar and Miguel Corte-Real in 1501-02, the markedly different fishing techniques required for trans-Atlantic fishing would have been difficult to adopt in the span of only a few years, suggesting that Portuguese fishermen were frequenting American waters before the Corte-Real brothers. Records in the Azores show, for example, that João Vaz Corte Real, father of Gaspar and Miguel, had already returned from a voyage to "Bacalhaos" by 1474.

92. And even, on the 1544 world map of Gemma Frisius, served to denote the entirety of North America. The term "bacallao" itself, in its various forms, is often thought to be of Basque origin, principally because it is the Basques who are commonly believed to have begun the dried cod industry, but in fact neither point is certain.

93. Legends translated according to Thacher, The Continent of America.

94. Possibly Jacobus Cnoyen or Nicolas of Lynn. See Mercator world map, entry 31.

95. See page 34 in entry 9 (Beccari) for Brasil Island.

malevolent creatures developed a rich mythology through medieval lore. Gog and Magog were traditionally imprisoned behind the Caspian gates by Alexander the Great; belief in the menace was so great that the level-headed Roger Bacon hoped that the study of geography might predict when, and from what direction, their onslaught might come in the days of the Antichrist so as better to prepare to defend against them. This threat was quite real to Columbus, who figured himself prominently into the events, believed by him to be close at hand, leading to the end of the world.[96] In the gulf formed by North America and the ominous land of Gog and Magog lies the Spanish Main, the Caribbean, which in Ruysch's mind was really the China Sea. Here the influence of Columbus is profound.

The Spanish Crown's acceptance of Columbus' proposal had rested on his dramatically reduced estimate of the distance across the western ocean to Asia. To achieve an acceptable figure, Columbus chose from various sources the most advantageous figure for each variable in the equation, and combined them. He took the widest longitudinal breadth for the known world to be found in classical sources, 225°, the figure proposed by Marinus of Tyre which had been recorded (and rejected) by Ptolemy. He then calculated the remaining 135° at the shortest value for a degree he could justify. Corroboration came from the apocryphal Second Book of Esdras, which stated that "six parts has thou dried up," interpreted by Columbus to mean that water covered only one seventh of the earth's surface. Columbus further reduced the expanse of ocean to be crossed by subtracting 30° for islands said to lie off the Asian coast. Those islands were, principally, Japan and Antilia.

Closer to Europe and first in his path was Antilia, mythical island of the Seven Cities. "Antilia" and the "Seven Cities" were originally two separate traditions.[97] The legend of the Seven Cities apparently originated in the Moorish conquest over Roderick, last king of the Visigoths, circa 711. It was said that seven Visigoth bishops and their people escaped to Portugal, from where they set to sea and discovered an island on which each bishop founded a city. This was a curious twist: when Columbus won the consent of Ferdinand and Isabella to undertake his voyage west, Spain had just fully rid itself of Moorish domination with the fall of Granada, and Columbus to an extent was riding the wave of Spain's new-found self-confidence and independence. Thus while Iberia's eventual freeing itself of Moorish clutches helped spark her urge to explore, a legend resulting from the initial Moorish conquest in Europe established a fantasy which those explorations sought. Just after the mid-fifteenth century the island of Antilia merged with the tradition of those Seven Cities.[98]

Both Japan and Antilia were described by the influential Italian humanist Paolo Toscanelli in his famous letter of 1474 to King Philip of Portugal, a copy of which Columbus read.[99] Ruysch places the island of Antilia tantalizingly close to the Azores, and accompanies it with an inscription summarizing its history:

> *"This island of Antilia was discovered by the Portuguese, and now when it is sought it is not found. In this island are people who speak the Spanish tongue, and who in the time of King Roderick are believed to have fled to this island from the barbarians who at that time invaded Spain. Here dwelt an archbishop with six other bishops, each one of whom had his own particular city. Wherefore this island is called by many 'seven cities.' This*

---

96. Columbus began to write (but never finished) a "Book of Prophecies" which was to record history and the part he was to play in it. Much of the inspiration for the parts of it which survive come from Pierre d'Ailly's opuscula, which included reflections about Gog and Magog. See Pauline Moffitt Watts, "Prophecy and Discovery: On the Spiritual Origins of Christopher Columbus's 'Enterprise of the Indies'," in The American Historical Review, Vol. 90, February, 1985.

97. See pages 33-34 in entry 9 (Beccari) for Antilia.

98. Cortesão claims to have deciphered "antilia septe civit" on the nearly-illegible Weimar/Freducci chart of circa 1460. See Imago Mundi X.

99. See introduction and footnote 7. In 1474, Toscanelli, responding (via an intermediary) to an inquiry from Afonso V of Portugal, stated that the Indies could be reached by sailing west and that the islands of Japan and Antilia lay en route.

> *people lived most piously in the full enjoyment of all the riches of this time."*

Later, not having been found in the Atlantic, the Seven Cities were sought on the American mainland, where they remained an elusive goal of riches-hungry Spanish explorers through much of the sixteenth century.[100]

In Ruysch's Caribbean there lies a strange triangular landmass, undefined on its western coast, in the position one would expect to find Cuba. But "Cuba" does not appear as such, and historians have long been puzzled by Ruysch's apparent omission of the island. Recent research, however, has revealed plate erasures under this land proving that Ruysch had in fact originally plotted a "true" Cuba, but that he subsequently altered the copperplate.[101] The vestige of *DE CVBA* is still discernable, as is part of the coastline of its original (insular?) geography. Preceding "de Cuba" was probably the word *Insula* or *Terra*. It can be surmised that Ruysch was confused by Columbus' insistence that Cuba was the eastern tip of a continental Asian promontory; Columbus, on his second voyage (1493-94), went so far as to force his officers to make sworn statements that Cuba was such a peninsula, in effect the *Aurea Chersonesus* or *Cattigara* of Ptolemy. Ruysch neatly covers his indecision on the matter with a mapmaker's "fig-leaf," a neat ribbon hiding its open end with an inscription stating that *"as far as this the ships of Ferdinand* [i.e., Columbus and the Spanish expeditions] *have come."* This became a popular device for mapmakers to avoid the unknown through the eighteenth century. The easternmost point of this land is *C. de Fundabril* (i.e., "end of April," probably present-day *Punta Maisi*), which cape Columbus had dubbed *Alpha and Omega* on his first voyage.[102] This first European name given to Cuba's most easterly point was meant to signify that it was both the beginning *(alpha)* of the Indies to those approaching it from the east, and the end *(omega)* of the Indies for those who had approached it from the west. Such is how Ruysch's *C. de Fundabril* would have appeared had his "Cuba" connected to the Asian mainland.

More havoc besets the island just east of pseudo-Cuba. This island lies between 45° and 55° east of the coast of Asia, just above the Tropic of Cancer. Ruysch's problem was that two of his prime sources of data —— the recent Spanish expeditions, and Marco Polo —— both located an island at that spot. For Columbus it was the island of Hispaniola (modern-day Haiti and the Dominican Republic). But according to Marco Polo, it was Japan *(Cipango)*, which the Venetian merchant described as lying *"far out to sea to the eastward* [from China], *some 1,500 miles from the mainland."*[103]

---

100. In 1536 Indians in the American Southwest spoke of Cibola, a Zuñi region, with seven large cities, and from this the legend of Antilia and its Seven Cities evolved into "The Seven Cities of Cibola." See page 98 in entry 29 (Ramusio).

101. Donald McGuirk located the erasure on an example of the first state (and hence earlier strike) of the map, though now that the erasure is known it can also be discerned on heavily-inked examples of later states, such as the present one. For a complete discussion of the Cuba and other reconstructions see: Donald McGuirk, "The Mystery of Cuba on the Ruysch Map," The Map Collector, no. 36 (September 1986); "Depiction of Cuba on the Ruysch World Map," Terrae Incognitae, Vol. XX, p. 89-987 (1988); and "Ruysch world map: census and commentary," Imago Mundi 41 (1989).

102. According to Ferdinand Columbus, las Casas, and Martyr, Columbus mentioned retrospectively having used this name on his first voyage, while speaking about the second voyage, although the name does not actually appear in the first voyage account.

103. Polo's 1,500 miles probably refer to Chinese miles ("li"), approximately one third of the Italian mile and thus remarkably close to the true distance separating China from Japan. Ruysch, of course, did not understand this. Ruysch's 1,500 (Italian) miles corresponds roughly to 45 to 55 degrees (far too much) because he underestimated the length of a degree. Ruysch's inscription also noted that Polo's description of Japan matched the Spanish description of Hispaniola, "except for the idolatry" of its people. Note that some translations of this Ruysch legend quote 1400 rather than 1500 miles, as Ruysch's "5"s have sometimes been mistaken for "4"s. Comparison of the number to latitude scale on the sides on the map

Hence Ruysch charts Hispaniola but accompanies it with a disclaimer stating that the island must in fact be Japan :

> *"M Polo says that 1500 miles to the east of the port of Zaiton there is a very large island named Sipagu . . . but as the islands discovered by the Spaniards occupy this spot, we do not dare to locate it here . . . being of the opinion that what the Spaniards call Spagnola is really Sipagu."* [104]

Below Antilia lies *LE XI MIL VIRGINE* ("[Islands of] the Eleven Thousand Virgins"), an early reference to the Virgin Islands named by Columbus for the story of St. Ursula. In this tradition, popular in the Middle Ages, King Dionotus of Cornwall had promised his daughter Ursula to the king of Brittany, a pagan. Wishing to preserve her virtue and enrich her life, the Christian Ursula solicited her father's consent for a three-year reprieve before her marriage, during which time she wished to travel to distant lands. She selected ten other noble virgin women to join her. But when word of the scheme spread, fully 10,988 other maidens from her father's kingdom elected to join the entourage, making a total of 10,999 virgins. The tally reached eleven thousand virgins when they reached Rome, for there the pope was so taken by the women's story that he abdicated his post and joined them. At the close of their alloted three years they were loath to return, however. Divine intervention produced unfavorable winds which postponed the sail back home and ultimately spared them a worldly fate, though tragically. Opting to visit Cologne when by misfortune Attila and the Huns were ransacking the city, they were raped and murdered. The story was a popular one for explorers seeking to name a complex of many small islands. Columbus, first sighting the many crests of the Virgin Islands, expected them to form one large island. When each peak proved to be a separate isle, the legend of St. Ursula seemed appropriate for naming them.

Another inherited tradition involving the "fairer sex" is found in the island of *MATININA* to the south. Beginning the homeward trek on his first voyage, Columbus claimed that the Indians told him of the location of this island, which was inhabited solely by women. Although concern with leaky caravels prevented their making a detour to visit the island, Columbus was

> *"certain that there is such an island, and that at a certain time of year men come to these women from the Isla de Caribe, which is 30 or 36 miles from us; if the women give birth to a boy they send him to the island of men, and if a girl they keep her with them."* [105]

Thus the tradition of a land of Amazonian women, long imbedded in European folklore, was transplanted to the New World.[106]

In the ocean between the triangular pseudo-Cuba and the Asian mainland a marked inconsistency in the stipling of the ocean is easily visible; this is because an inscription had first been engraved there, which, like "true" Cuba, was then pounded from the copperplate. Very faint vestiges of the original inscription are discernable, however, and painstaking reconstruction has recently resurrected it from oblivion. The legend which Ruysch abandoned was a quote from the *Imago Mundi* of Pierre d'Ailly (circa 1483), one of Columbus' favorite texts, describing the people of the far northern latitudes according to Pliny and Martian. It stated that :

(graduated by 5 degree increments) shows that Ruysch did in fact write 1500, rather than his having consulted a variant copy of Polo which might have said 1400.

104. In a fascinating example of how a geographic misunderstanding can assume a life of its own, Donald McGuirk has demonstrated how Ruysch's hybrid Hispaniola-Japan was transformed into an atypical representation of Japan four years later by Bernard Sylvanus. Columbus himself wrote in his log that he believed Cuba to be Japan (October 23, 1492). See IMCoS Journal Vol. 4, no. 3, p. 5-9 (August, 1984), and entry 14 in this volume.

105. First voyage log, January 16, 1493.

106. See, e.g., the "mare amasoneorum" on entry 2 (color plate I), located on that map at the far north (9 o'clock position).

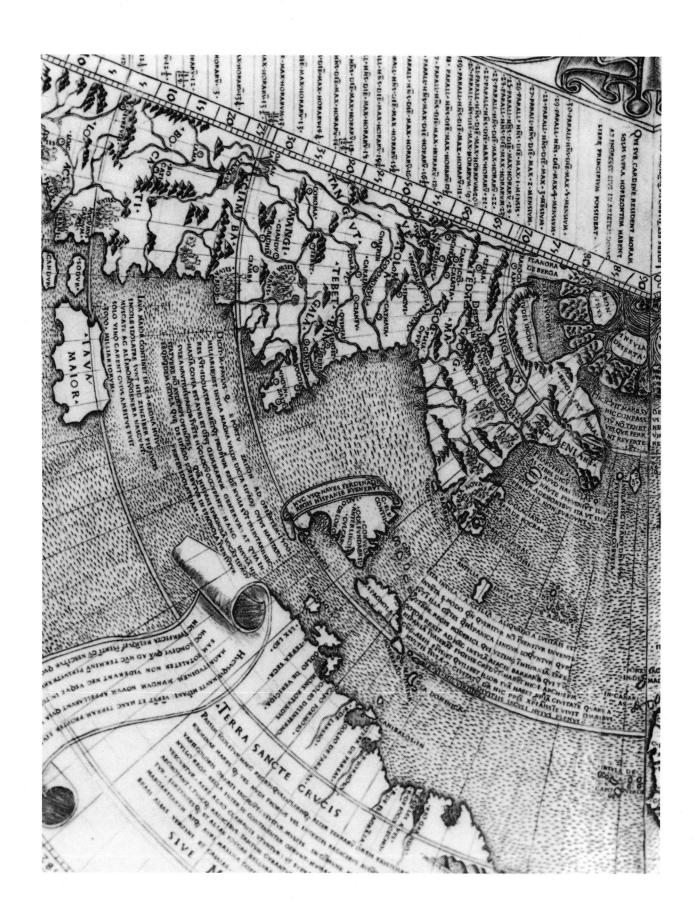

PLATE 8   Johann Ruysch, Rome, 1507 (detail)   (entry 12)

"how far the habitation extends north Pliny shows through actual experience and by various authors. For habitation continues up to that locality where the poles are located; and where the day lasts six months and the night for the same length of time. Martian, moreover, in his description of the world, agrees with this statement; whence they maintain that in those regions dwells a very happy race, which dies only from satiety of life, attaining it casts itself from a lofty rock into the sea. These people are called Hyperboreans on the European side and Arumphei in Asia." [107]

South America appears as a large and distinct continent labelled *TERRA SANCTE CRUCIS*. This "Land of the Holy Cross" contains the term *MUNDUS NOVUS*, the first reference to a "New World" on a printed map.[108] South America is mapped through the Gulf of Venezuela, with the Guahira Peninsula shown as an island *(TAMARAQUA)*. No trace of Columbus' fourth voyage (1502-03), in which he stumbled across Central America, is found on Ruysch's map.

On South America's northern coast is the *GOLFO DE PAREAS* which Columbus believed lay at the foothills of Paradise. Columbus reached this fancy when attempting to explain the variation of the North Star, noted by him near the Gulf of Paria. Peter Martyr shared the Admiral's bewilderment, declaring

"but how it commeth to passe, that at the beginnynge of the evenyng twilight, it is elevate in that Region only fyve degrees in the moneth of June, and in the morninge twylight to bee elevate xv. degrees by the same quaadrante, I doo not understande." [109]

Probably inspired by d'Ailly, Columbus here reached his famous conclusion that the variation occured because he was actually sailing "uphill" and that in that place, *"near the equator in the Ocean Sea at the end of the Orient,"* the Earth took the shape of the nipple of a woman's breast. At the pinnacle of this nipple was the part of the Earth which *"was the highest and nearest to heaven."* D'Ailly's text, of which Columbus was very fond, relates that Paradise *"is so elevated that it touches the Lunary Sphere"* and that *"its altitude relative to the level of the low earth is incomparable."* So, Columbus adopted myth to the scene. *"In soo muche that he ernestly contendeth,"* states Martyr, *"the earthly Paradyse to bee situate in the toppes of those three hylles."*

Columbus' association of Paria with Paradise was reinforced by the fact that there he encountered four rivers emptying into the gulf, corresponding to the number of rivers commonly believed to flow from Paradise. D'Ailly had declared that there was *"a fountain in the Terrestrial Paradise which waters the Garden of Delights and which flows out by four rivers,"* and Columbus himself annotated his copy of the *Imago Mundi* with *"Fons est in paradiso"* ("a fountain is in Paradise").[110] And such is how he construed what Martyr described as *"the outragious streames of the freshe waters,"* being the Orinoco River's discharge, *"whiche soo violentlye isshewe owtt of the sayde govlfes and stryne soo with the salte water, saule headlonge from the toppes of the sayde mountaynes."*

Nearly blocking the entrance to Columbus' fountainhead of Paradise, the Gulf of Paria, is *CANIBALOS IN* (Trinidad), introducing to maps the cannibalism which Columbus reported

---

107. As with the Cuba erasure, this reconstruction was the work of Donald McGuirk. Gregory McIntosh noticed the correlation to the Imago Mundi following McGuirk's near-complete reconstruction. D'Ailly himself took the words from Roger Bacon's Opus Majus of circa 1267. See "Ruysch World Map: Census and Commentary" in Imago Mundi 31, p. 137-141 (from which the d'Ailly legend is quoted).

108. A world map by Francesco Rosselli, tentatively dated as "circa 1508" by Shirley but which could possibly be as early as 1506, contains the term "Mudus Novus."

109. This and following Martyr quotes from Martyr/Eden Decades, p. 32

110. See the typical representation of Paradise and its four rivers in entry #2, Le Rouge. D'Ailly quote from Polk, The Island of California.

about the Carib islanders. It is interesting to note that on State I of the Ruysch map *Canibali* designated the island of Dominica *(LA DOMINICA)* instead, which was one of the islands Columbus said was inhabited by that reputably fierce race. While at Dominica on the Second Voyage, a colleague of Columbus wrote that *"these islands are inhabited by Canabilli, a wild, unconquered race which feeds on human flesh. I would be right to call them anthropophagi."*[111] That Ruysch, on the later states of his map (such as the present example), chose to relocate the cannibalism reference close to South America may reflect early skepticism about Columbus' reports of the Caribs' supposed eating of human flesh. The influential Las Casas, for example, stated flatly that the Caribs were not cannibals. Cannibalism had, though, been reported independently by early mainland explorers, such as Vespucci, perhaps rendering the Trindad association more plausible in Ruysch's mind.

On South America, a long inscription informs the viewer that this is the

> *"Land of the Holy Cross or the New World. This country, which is generally considered another continent, is inhabited in scattered settlements. The men and women go either entirely naked or adorned with interwoven fibers of wood [i.e., roots] and birds' feathers of various colors. Many live in common: they have no religion and no king: they wage war among themselves continually. They devour flesh —— that of captives in war: they breathe so mild an air that they live to be over 150 years: they are rarely sick, and then they are cured by roots or herbs only: lions are found here: serpents and other foul beasts. There are forests, mountains and rivers: there is the greatest abundance of pearls and gold. Brazilwood, otherwise called verzini, and cinnamon are exported hence by the Portuguese."*

The nomenclature along the Brazilian coast further reflects Vespucci's adventures. The atypically colorful reference to an *abbey* in *ABATIA ONIV SACTORV* ("All Saints' Abbey") is found both in the original Italian printing of his fourth voyage account (as *badia*) and in translations (e.g., *abbaciam* in the *Cosmographiae Introductio*) but is probably a corruption of *baia* or *bahia* (bay).[112] This site had been a welcome find for Vespucci, who, having followed the course of his previous voyage south along the coast, *"at last found a harbor which we called Bay ('Abbey') of All Saints."*

Ruysch's *CAPVT S. CRVCIS* is the Cape St. Vincent of Vespucci's third voyage (1501-02), hence the name of the mountains shown just inland by Ruysch. Vespucci said this cape lies eight degrees south of the equator (modern-day Recife), and that thence sailing southwest along the coast they encountered friendly Indians *"gazing in wonder at us and at the great size of our ships."* Vespucci remarked that a few of the Indians, taken *"to teach us their tongue,"* volunteered to return to Portugal with them. At the southern end of South America another inscription states that the Portuguese had sailed as far south as 50° latitude without seeing the southern limit of the land, possibly also a reference to Vespucci's third voyage. On that voyage he claimed to have sailed so far south that

> *"the South Pole rose fifty-two degrees above the horizon, and we could no longer see the stars of the Great or the Lesser Bear [and] on the 7th of April, when the sun was near the end of Aires, we found that the night was fifteen hours long."*

---

111. Extracted from a letter of Guillermo Coma of Aragón which was translated into Latin and published in late 1494 or early 1495. Columbus' own journal from the Second Voyage does not survive. Cannibalism may have been a prefered scenario because only it, and the refusal of conversion, was condoned by the Crown as grounds for enslavement of the Indians. See Sauer, <u>The Early Spanish Main</u>. Translation quoted from Kirkpatrick Sale, <u>The Conquest of Paradise</u>, p. 129.

112. This and the following Vespucci quotes are from Waldseemüller, <u>Cosmographiae Introductio</u>, 1507. As regards the circumstances, remember that the authenticity of Vespucci's fourth voyage is not established.

The western bounds of the continent remained entirely unknown. In 1507 the Pacific, still simply the "Oriental" or "China" Sea, had never been sighted from the American side and had only been cursorily approached from the Asian side. Ruysch places a ribbon over the west coast of South America with an inscription stating that

> *"Thus far the Spanish sailors have come, and because of its magnitude they call it a new world, for indeed they have not seen the whole of it nor at this time have they explored beyond this limit. Therefore this map is left incomplete for the present, since we do not know in which direction it trends."*

The Portuguese, of course, had also "come this far." Their discovery of Brazil in 1500 was significant because it was rich in trade potential and because it clearly lay on their side of the papal demarcation line.[113]

During this excitement in exploration to the west, Portugal continued pressing forward around Africa, establishing regular trade with India, their other papal-sanctioned sphere of influence. As a result of Ruysch's close contact with Portuguese sources, he shows for the first time the true relative proportions of Sri Lanka and the Indian subcontinent.[114] Further east, however, new data has become entwined with old. It is here that Ruysch betrays his reliance on the then two-hundred year old reports of Marco Polo.

Some of Ruysch's inland Asian data was extrapolated from Polo's description of his trip from Peking to Bengal, which he says he made as an emissary for Kublai Khan. Ruysch judges Polo's *TOLMA*[N], a region later reincarnated in the New World as part of the Northwest Coast, to lie just south of the realm of Gog and Magog. South of *TOLMA* lies the mountainous country of *TEBET* (Tibet, but actually in present-day Sze-ch'wan and Yün-nan), which *"is terribly devastated, for it was ravaged in a campaign by Mongu Khan . . . many towns and villages and hamlets lying ruined and desolate."* Polo did, however, *"renew his stock of provisions"* at a region with many populated hamlets, some of which are *"perched on precipitous crags."* At these villages he observed such promiscuous social customs that *"obviously the country is a fine one to visit for a lad from sixteen to twenty-four."*

Most of Ruysch's Asian data comes from Polo's description of his return to the West from China.[115] Marco, Maffeo, and Niccoló Polo, long detained at the pleasure of Kublai Khan, were allowed to return to Europe when sent as the personal escorts of a bride for Arghun, khan of the Levant (Persia).[116] They departed China from *Zaiton,*[117] seen as a

---

113. See footnotes 82 and 144.

114. With the possible exception of the Waldseemüller map "Orbis Typus Universalis," (entry 11), not known to have been published until 1513. Some of the geographic data brought back by the Portuguese voyages to India probably came from the Arabic or Javanese pilots who helped guide her vessels in the Indian Ocean.

115. Polo's reports in his book The Travels can be roughly divided into three categories of accuracy: those that came directly from his own observation, which have proven themselves to be reliable and non-speculative (e.g., a Chinese document discovered earlier this century actually corroborates the names given by Polo of the leaders of this Zaiton-Persia voyage) ; those that he himself reported, but from heresay rather than from direct observation (such as an Arabian Nights inspired tale of birds on Madagascar which carry elephants away) ; and pure extraneous embellishments (such as formula descriptions of battles) which appear to have been added by his writer Rustichello to embellish sections dealing with the less "exotic" Near East, rather than the Far East.

116. Queen Bulagan, wife of King Arghun of Persia, had died. As she stipulated in her will that only a woman of her own ancestry should sit upon her throne, Arghun sent envoys on horseback to Kublai Khan to request a suitable relative to be his new queen. A woman of seventeen named Kokachin was chosen and sent back, again via land, to Arghun. War between Tartar kings had recently broken out, however, rendering the roads impassable, and so they returned to China after a futile eight months of travel. Thus obliged to use the sea route instead, the Persian envoys asked that the three Polos, whom they described as "experienced Latins," be allowed to accompany them.

coastal city on the Asian peninsula opposite Ruysch's "Cuba." From Zaiton they travelled *"1500 miles"* across a gulf to a country called *Chamba*, which is plotted here slightly inland as *CIAMBA*. Ruysch designates its coast as *SILVA ALOE* (aloe forest) for the valuable plant which Polo said the king of *Chamba* offered as part of his annual tribute to the Kublai Khan. Groves of trees yielding a black wood, which Polo said was used for making chess-men and pen-cases, are noted by Ruysch as *SILVA EBANI* (ebany forest). To the west of these forests lies *LOAC* (Thailand/Malaya). Sailing seven hundred miles south-south-west from *Chamba* the traveller, says Polo, passes two uninhabited islands, Ruysch's *SODVR* and *CANDVR*. Among the Polean islands lying further to the south on Ruysch's map is *AGAMA*, which like *TOLMA* would later be transposed to the New World as a region of the Northwest Coast.[118]

With the khan's entourage, Polo followed the lengthy sea route into the Indian Ocean through the Malay Straits between the Malay Peninsula and Sumatra, learning of (but certainly not visting) Java. After passing five months on the northern coast of Sumatra waiting for favorable sailing weather, they continued to Sri Lanka and India, ultimately arriving at Hormuz, from where the Polos, having completed their mission, made their way back to Venice. Mixing Polean, Portuguese, and Ptolemaic data, Ruysch charts a *IAVA MAIOR* and *IAVA MINOR,* Polo's Java and Sumatra respectively. He creates a "modern" Sri Lanka *(PRILAM)* from Portuguese sources but exiles the old *TAPROBANA*, the Sri Lanka or Ceylon of Ptolemy, to the east, giving it both the old and new names *(TAPROBANA ALIAS ZOILON)*, and mistakenly notes the 1507 Portuguese landfall in Sri Lanka by it rather than by *Prilam*. Further east is yet another Sri Lanka *(SEYLAN INSULAE)*, spanning the edges of the map.[119] South of *IAVA MINOR* (Sumatra) an inscription refers to an archipelago of precisely 7,448 islands reported by Polo, probably the Philippines but here suggesting Indonesia.[120]

The plate re-working mentioned above in regard to Cuba is the most significant of many changes made to the two plates from which the map was printed both before and during its publication. Of the three states of each half known to exist, the present example is the last state on both sides, as are almost all extant copies.

The extent and manner in which this map underwent modification prior to, and during, the approximately one year it was published, provides a rare insight into the inner conflicts, uncertainties, and frustrations that makers of serious Renaissance maps experienced in their attempts to reconcile the new and distant discoveries with the world as it had been known. It also demonstrates that Ruysch's map was not an exact derivative of any one prototype, but rather that he was grappling with diverse sources according to his own judgement. Ruysch compounded his difficulties by ambitiously opting to chart the full 360° breadth of the earth's longitude, for which task he adopted the fan-shaped projection used the previous year by Contarini. This early, but short-lived, solution to the problem of representing the sphere on a plane surface was essentially an outgrowth of the conical (first) projection of Ptolemy. Contarini and Ruysch further "unfurled" Ptolemy's "fan" so that the two "ends" of the map "meet", that is, they represent the same meridian. That meridian, in turn, is the continuation around the other side of the world of the map's center meridian, the one which skims the west coast of Africa. The Ptolemaic stage has also been lengthened "down" from its pivot (the North Pole) to include new discoveries south through beyond the Tropic of Capricorn.[121]

Ruysch's map was normally part of 1508 issues of the "Rome" edition of Ptolemy, previously published in 1478, 1490, and 1507. The map has, however, also been found in 1507

---

117. Zaiton is probably Chüan-chau, but certainly the Amoy harbor region.

118. See entry 33 (Cartaro globe), and footnote 253.

119. Ptolemy's "Taprobana" has, in fact, sometimes mistakenly believed to have represented Sumatra, as it appears to be in this incarnation despite the "modern" Ceylon name.

120. See footnote 211.

121. In mapping the full east-west span of the earth's surface, the Ruysch is preceded by the map of Contarini-Rosselli (1506); neither of these maps, however, charts latitude fully to the south pole. The first maps to show the entirety of the earth's surface are the Waldseemüller globe gores of 1507 and the oval Rosselli map of circa 1508.

issues of the atlas with the original binding intact (and hence not a later addition to the volume), and so the map is assigned a date of 1507. The fact that it is not normally found in the 1507 issues of the atlas, and especially the map's noting of an event of 1507 (the Portuguese in Ceylon), indicate that the map was probably readied for publication only at the very end of that year. This immediate assimilation of advanced (and probably confidential) Portuguese data in the Indian Ocean also demonstrates that Ruysch must have had very direct taps on Lusitanian intelligence.

- - - - - - - - - - - - - -

13. THE WORLD

Tipus Orbis Universalis IVXTA Ptolomei Cosmographi Traditionem et Americi Vespucci Aliorque Lustrationes a Petro Apiano Leysnico Elubrat. An. Do. 1520.

[A Map of the World based on Ptolemy and Amerigo Vespucci by Petrus Apianus, the Year of Our Lord 1520]. Vienna, 1520. World map constructed on a truncated cordiform projection, from Caius Julius Solinus' *Ionnis Camertis minoritani . . . ennarationes.*

Medium : woodcut.
Size of original : 285 x 410 mm.

PLATE : 9

In 1507, the year following Columbus' death, a group of humanists met in St. Die under the patronage of the Duke of Lorraine. Their goal was to derive a working hypothesis from the myriad new and often conflicting geographic data that had been compiled from the voyages of the preceding fifteen years. Most notable among the scholars, engravers, printers, and draftsmen was the eminent geographer Martin Waldseemüller.

The major fruit of the enterprise, known as the Gymnasium Vosagense, was a radical new map of the world on twelve woodcut sheets. This work gave the New World its own turf and its own name. Here, for the first time, the New World took its place as a distinct and autonomous part of the globe, with North and South America forming a nearly contiguous continent partitioning into two the ocean separating Europe from Asia. Hence the Pacific Ocean as a *concept* has now been formally added to the world map. And here, for the first time, the New World (South America) is called *America*.[122]

In modern times, this map of Waldseemüller was known only through inference until the only extant copy was discovered in 1901. We do not know if the map had been widely

---

122. This map, entitled "Universalis Cosmographia . . .", is assigned the date of 1507, as is a set of Waldseemüller's gores for a small globe. Both the large map and the gores follow the same geography, except that the gores contain only minimal detail, and do not truncate South America. It is possible, however, that the sole surviving example of the large map was not actually printed until after 1515, and that only the gores (of which two sets survive) were printed in 1507. But as the geography and use of the name "America" clearly originated in 1507, the relative dating of the printed impressions is not important here. (See Elizabeth Harris, "The Waldseemüeller world map: A typographical appraisal," <u>Imago Mundi</u>, volume 37, 1985). See also footnote 85.

proliferated.[123] But even if little-known in its own day, its influence was nonetheless secured through this adaptation done thirteen years later by Peter Apianus.

Apianus took the 1507 map and lowered the truncation of the cordiform to allow his charting of a hypothetical configuration for the southern extreme of the New World.[124] This was significant, because finding a passageway around (or through) America was at this time of the utmost urgency; it was in fact at precisely the time Apianus made this map that Ferdinand Magellan was en route south along the South American coast in search of such a passage (the expedition left in 1519 and reached the strait named after him in October of 1520).

In all probability Magellan and his crew expected the continent to end at about the point Apianus has shown. This was partly because that point parallelled the southern coast of Africa and thus satisfied classical belief in a balanced earth, and also because of early reports of the wide gulf of the Rio de la Plata. Both the northern shore of the estuary of the Plata and the southernmost tip of Africa occupy uncannily close parallels, both lying at just shy of 35° south latitude. Further, the Plata's mouth is an astounding 190 km wide and maintains a wide profile for over 100 km upriver, thus appearing as a sea rather than the entrance to a river. This amazing three-fold coincidence would have reinforced an assumption by pioneering explorers, reaching the Plata, that they had reached the southern end of the continent. In reality, as Magellan was woefully learning, America extends more than fifteen hundred km further south.[125]

Apianus' North America is copied without emendation from the 1507 Waldseemüller map. It is depicted as a large, north-south oriented island labelled simply as the "unknown land on the other side" *(Vlteris Terra incognita)*. Opposite Cuba *(Isabella)* is what appears to be the Gulf of Mexico and Florida Peninsula, although the latter may instead be a vestige of a peninsular Cuba.[126] Like Waldseemüller, Apianus mistakenly locates *Parias,* a gulf and region of the Venezuela coast first described by Columbus, on the southern end of North America.[127] Its transposition to North America may have been an attempt by Waldseemüller to separate Columbus' discoveries from those of Vespucci.

Apianus has also retained Waldseemüller's use of the term *America* in recognition of the supposed discoveries of Amerigo Vespucci.[128] Much mystery and controversy surrounds Vespucci and his travels to the New World. Waldseemüller's decision to bestow his name upon the New World rested on his acceptance of a voyage purportedly made by Vespucci in 1497. Vespucci supposedly reached the South American mainland on that voyage, thereby winning precedence over Columbus, who did not actually reach the mainland until 1498. One school of

---

123. Although only one example, luckily preserved in a book which had belonged to Johann Schöner, is extant, records indicate that 1,000 copies were printed. Such a poor survival rate is possible with a large, separately issued map.

124. The pattern used by Apianus is also found in a primitive fashion on the 1507 globe gores of Waldseemüller.

125. The recorded discovery of the Plata was by Juan de Solís in 1516 (who referred to it as a "fresh sea"), but news of it probably reached Europe earlier from other voyages. The Ruysch map of 1507, for example, records Portuguese progress to about that point (see pages 48-49). Magellan may have seen Schöner's globe of 1515, which charts a strait between South America and a southern continent at about 45 degrees south latitude, possibly based on an account of a Portuguese voyage to the Rio de la Plata contained in a German newsletter entitled Copia der Newen Zeytung ausz Presillg Landt. See pages 71-72 in entry 19 (Finaeus).

126. No European encounter with Florida is proven prior to Ponce de León in 1513. See entry 15 (Martyr) and 16 (Waldseemüller).

127. On the larger and more detailed 1507 Waldseemüller map, Parias is even correctly depicted as a mammoth river delta as per Columbus' description from his third voyage (1498).

128. Lorenz Fries, who cut the woodblock for Apianus' map, was probably instrumental in the ultimate acceptance of the term "America." Two years after this map, Fries made another world map using the term, the first to appear in an atlas; that 1522 map appeared in a popular edition of Ptolemy which underwent four editions through 1541.

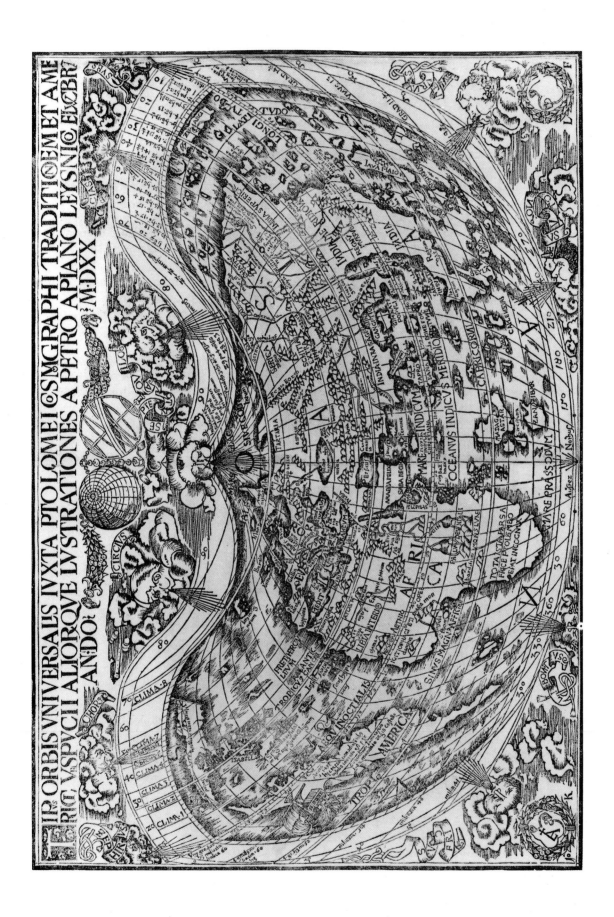

PLATE 9   Martin Waldseemüller /Peter Apianus, Vienna, 1520   (entry 13)

thought holds that Vespucci fabricated this first (historically most critical) and his fourth voyages, and that he at least exaggerated his role in the other two. Even early chroniclers, notably Las Casas and Herrera, accused Vespucci of fraud, of conniving to secure his own immortality and rob Columbus of his due fame. The counter-argument holds that although Vespucci's 1497 voyage was indeed ficticious, he himself was quite innocent of the whole affair.

As a Sevillian representative of a firm based in his native Florence, Vespucci helped supply ships for, and took part in, Hojeda's expedition of 1499-1500. This was the second of the four voyages, the first of his two "proven" voyages. On this voyage they explored the northern coast of South America, checked for possible English advance from the north and, notably, attempted to use astronomical observations to approximate their longitude. In 1501 he returned to the New World, this time in the employ of Portugal and charged with exploring the coast of Brazil to determine the extent of land which lay within that country's side of the line of demarcation. Vespucci wrote about these two voyages to a friend in Florence, Lorenzo di Pier Francesco de Medici. In 1503, with Lorenzo perhaps deceased, these letters came into the hands of a publisher who circulated them in printed form. Late the following year there appeared another letter purportedly by Vespucci, this one addressed to Piero Soderini, who had been elected head of the Republic in 1502 and who was soon to become mentor in military affairs to Machiavelli. The new letter added two voyages to Vespucci's repertoire, of which the first in particular, said to have been made for Spain in 1497, is suspect. The letter, written in language not typical of an educated Florentine (which Vespucci most certainly was), was quite possibly a fabrication of the printer. If true, then Vespucci was not culpable for the immortality bestowed upon him in St. Die, as it was this Soderini letter (in French translation) which motivated Waldseemüller and his colleagues to invent the term *America*. This sort of license and fraud was a common problem in these early days of printing, and one of which many writers complained. Additionally, the infamous rivalry between Italian city-states may have prompted Soderini, like Vespucci a Florentine, to seize the opportunity to champion his city over Columbus' Genoa. Vespucci himself had become a citizen of Spain in 1505. In 1508 he was appointed the first "pilot major" of the *Casa de la Contratación de las Indias*, an arm of the Crown which had been established in 1503 to carry out its operations in the "Indies."[129] Vespucci supervised the compiling of the *padrón real* (official government chart), a task which he pursued until his death in 1512. Nothing in his activities suggests that he sought glory, and there is no incriminating evidence that he himself ever misrepresented his adventures. There is not, in fact, any indication that Vespucci was ever even *aware* of this colossal quirk of posterity.[130]

Waldseemüller, writing in the *Cosmographiae Introductio,* a booklet meant to accompany his 1507 map, explains his decision to christen the New World with Vespucci's name:

> " [A] *fourth part* [of the earth] *has been discovered by Amerigo Vespucci . . . inasmuch as both Europe and Asia received their names from women, I see no reason why anyone should justly object to calling this part Amerige, i.e., the land of Amerigo, or America."*

Next in his book, Waldseemüller quotes Dionysius to state that there is but one ocean, which takes various names as partitioned by continents and islands. Apianus added this ancient idea of an encircling "Ocean Sea," typified by the "T-O" map of Isidorus, to his

---

129. The "Casa" controlled the licensing of pilots, collected and safeguarded geographic data, and inspected and taxed all vessels engaged in commerce with America. It remained in Seville until 1718.

130. It is interesting to note that Amerigo's nephew Juan, on his extant portolan of 1526, does not use the term "America" even though the seeds of its immortality had already been sown in published maps by Apianus and Fries, and Juan could have only benefitted from the association.

version of Waldseemüller's map by poetically placing the letters "O-C-E-A-N-U-S" around the perimeter of the cordiform.[131]

The Japan of Marco Polo, which had made its debut on a printed map one year before Waldseemüller's 1507 prototype, appears on Apianus' map a mere 10° west of North America (on the opposite end of the cordiform).[132] It is called *Zipargi* (Zipangri) after Polo's corruption of the Chinese *Jih-pên kuo*, meaning "land of the rising sun."

Waldseemüller's map integrates beauty and function. Its dramatic visual impact is a result of its imaginative cordiform projection, an early attempt to solve the troublesome new problem of mapping the entire breadth of a spherical earth onto a flat surface. With this heart-shaped projection, longitude becomes increasingly exaggerated away from the central meridian, as the outer "skin" of the map has been stretched to incorporate the "far" side of the earth. The projection also attempted to maintain equal scale throughout. It is interesting to note that the meridian Waldseemüller chose as that most convenient to split the earth for the cordiform, in the "Pacific" just east of Japan, is the line still considered the least disuptive to separate east from west : although Waldseemüller's Pacific is hypothetical and naively abbreviated, his "split" lies in the region of what is now the international date line.

Three types of cordiform projections were developed in the early sixteenth century. The present method is called a "truncated" cordiform because it does not extend to the south pole. Another, "true" cordiform projection, more successfully maintains equidistance from the north pole and charts the full north-south breadth of the earth, but appears to "compress" the features (e.g., Finaeus, entry 20). Lastly, a "double" cordiform (e.g., Finaeus, entry 19) divides the earth into northern and southern hemispheres, allowing a superior glimpse at the Arctic and Antarctic regions and minimizing distortion away from the poles, but losing continuity along the equator.

The idea of the cordiform itself had been used on a more primitive level by Henricus Martellus to depict the post-Dias, pre-Columbian world known to him. The heart shape already existed as a motif; it had been used poetically, for example, for musical notation in the fifteenth century.[133]

- - - - - - - - - - - - - - -

14. THE WORLD

[untitled world map on a pseudo-cordiform projection].

Bernard Sylvanus, Venice, 1511.
From Jacob Pentium's edition of Ptolemy's *Geographia*.

Medium : woodcut, printed in black and red.
Size of original : 565 x 415 mm.

COLOR PLATE : VIII

Although Vasco da Gama's success in reaching India via southern Africa in 1497-99 vindicated Portugal's earlier decision to abandon intensive exploration west of the Azores, the incentive

---

131. This was also added to later states of the earlier Ruysch map (entry 12).

132. The first appearance of Japan on a printed map was the Contarini-Rosselli world map of 1506. See also entry 29 (Gastaldi/Ramusio).

133. The exquisite "Chansonnier Cordiforme," made by an unknown scribe in Savoy for Jean de Montchenu before 1477.

to sail west had remained strong and some adventuresome individuals undertook, often at their own expense, further voyages of discovery into the western ocean. Little is known about most of them, and what we do know is generally restricted to those for which a privilege was first obtained from the Portuguese Crown.

The most influential was that of an Azorean land-owner by the name of Gaspar Corte-Real. Surely familiar with the latest in Azorean exploratory gossip, in 1501 he and his brother Miguel set out from Lisbon with a fleet of three ships and reached Newfoundland. Miguel returned to Portugal first, carrying with him many Indians as proof of their success, while Gaspar delayed his return to investigate further the new land. Gaspar, who probably entered the gulf of the St. Lawrence River, never returned, and Miguel, who then returned to America to search for him, vanished as well.

Sylvanus records the North American landfall of the Corte-Real brothers as *regalis domus,* a curious term in that it is "corte-real" inverted and Latinized. While endowing it with a politically neutral name, Sylvanus' placement of the land far to the east of the *raya* nonetheless secures it for Portugal anyway. Sylvanus may have been modifying a Portuguese prototype in the interest of discretion, as the map was published in Venice, adopted city of John Cabot, who had discovered land in the same region as the Corte-Reals for the king of England. In any event he is ambiguous about the land's nature, leaving it undefined on the west and omitting fully the 40° of longitude which would have revealed its relation to Asia or *Terra Sanctae Crucis* (South America). This suggests that Sylvanus was skeptical of the opinion expressed by the Venetian ambassador to Lisbon, Pietro Pasqualigo, that the new land extended south to the *"parrot land"* of Cabral (Brazil). Pasqualigo was present when Miguel Corte-Real's ships returned. He described the people brought back by Miguel, probably the Beothuk Indians, now long extinct. Visiting the first of the returning vessels, he related that:

> *"They say this country is very populous and the houses of its inhabitants are of long strips of wood covered with the skins of fish. They have brought back seven natives, men, women, and children, and in the other caravel which is expected from hour to hour, fifty others are coming. They resemble gypsies in color, aspect, figure, and stature. They are clothed in the skins of various animals, mainly of otters . . . They are very shy and gentle, but excellently formed as to arms and legs and shoulders, beyond describing."* [134]

Another noteworthy observer to the affair was the Italian Alberto Cantino, perhaps best described as a foreign spy.[135] Visiting the second ship, Cantino described Miguel's Indian captives as gentle people who laughed easily, showed their pleasure freely, and whose language was smooth sounding but not understandable. Specifically of the women, Cantino noted that :

> *"They have small breasts and most beautiful bodies and are of gentle countenance. Their color may be described as white rather than otherwise, the men being much darker. In sum, except for the guarded looks of the men, they are similar in every other way to ourselves as to features and*

---

134. Quoted from Sauer, Northern Mists, p. 50.

135. Cantino was in the employ of Ercole d'Este, Duke of Ferrara. Sometime after the Corte-Reals' return in 1501, and before the end of 1502, Cantino, back in Italy, presented a portolan chart of the world to d'Este. This map revealed the most sensitive and highly secret geographic data, up to and including that just gotten from the Corte-Reals. It is the earliest surviving chart to show the Portuguese discoveries in the New World and Indian Ocean (through da Gama, Cabral, and Corte-Real), and possibly the first to show Spanish discoveries in the New World (depending on the true date of the "1500" de la Cosa map). It was largely the basis for many subsequent maps. Evidence of Spanish sources is found in the spelling of place-names in the southeast of "Asia," possibly the Florida peninsula. This splendid map is in the Biblioteca Estense, Modena. That d'Este had a long-standing interest in strategic intelligence is evident from the fact that in 1494 he had visited Florence seeking Paolo Toscanelli's writings and maps.

appearance. They go quite naked except for a skin of the deer about their private parts. They have no arms nor any iron, but whatever they fashion and make they do by using very hard and sharp stone, with which they can work anything however hard." [136]

East of the homeland of these amiable people is *terra laboratoruz,* or Labrador, discovered by João Fernandes, a Portuguese-Azorian who sailed under the British flag and whose important voyage is contemporary with that of the Corte-Reals. Fernandes may have reached eastern Canada; Sylvanus' map, however, is one of several early works which contribute to the opinion that Fernandes actually reached Greenland instead.[137]

In South America, called *Terra Sanctae Crucis* ("Land of the Holy Cross"), the only place-name is *canibaluz romon.* The latter word is probably a mis-spelling for "domon," or dominion, i.e., "the domain of cannibalism."

In the Caribbean, just west of Cuba and Hispaniola (but on the other end of the map) lies Japan *(Zampagu Ins).* To find a pattern for his map's Japan, Sylvanus apparently looked to the Ruysch world map of four years earlier. Ruysch did not show Japan, but rather explained in an inscription that he believed Japan to be the same as one of the islands recently discovered by the Spanish (in the Caribbean). Although Ruysch indicated Hispaniola as that which he believed to be Japan, he placed an unidentified island —— or part of a northern promontory of a continental Cuba as per Columbus —— to the west.[138] For Sylvanus, it was much more sensible to adopt that homeless island as Japan, rather than Hispaniola, because it as yet had no identity, and because it was in the position of Cuba, which Columbus himself had once said was really Japan. Thus Sylvanus transplanted Ruysch's uncommitted "Cuba" to be "true" Japan, salvaging it from Ruysch's confused Caribbean.[139]

Sylvanus left the eastern shores of Asia and the western bounds of the Corte-Real's North America open, thus allowing for the possibility that they were contiguous. Although he also leaves the north coast of Japan undefined, his designation of that land as *ins*[ula] seems to make clear that he did not envision it connecting to the Asian mainland. No ambiguity whatever is found with Greenland *(GRVENLANT),* which Sylvanus integrates fully into the Asian continent, placing it due north of Cathay *(CATAI REGIO).*[140]

The rest of Asia speaks poorly for Sylvanus' sources. The East Indian islands are the by-now-familiar reconstruction of Marco Polo's voyage from China into the Indian Ocean, and they lie off the "false" Southeast Asia peninsula, a remnant of Ptolemy's Africa-Asia landbridge. But most disappointing is Sylvanus' apparent ignorance of data gained from the Portuguese voyages to India, as he still reverses the relative proportions of India and Ceylon.

This map, and the other maps in the 1511 *Ptolemy* atlas, were the first to be actually *printed* in color. Most examples of the map (the present one included) have a curious flaw: in Asia, the place-name *come,* meant as a black ink word, printed lightly in red as well. This quirk has helped reconstruct the method used to print the map, demonstrating that separate registers printed each color. The practice of printing maps in color was not pursued, however, and except for a few isolated attempts the printing of maps in color did not become established until the development of lithographic processes in the nineteenth century.[141]

- - - - - - - - - - - - - -

---

136. Quoted from Sauer, <u>Northern Mists</u>, p. 51.

137. See entry 18 (Bordone), pages 66-67.

138. See entry 12 (Ruysch), page 44.

139. See Donald McGuirk, "The New World in an Old Cage," <u>IMCoS Journal</u>, Vol. 4 no. 3, p. 5-9 (1984).

140. See footnote 191 for Cathay.

141. A map of Lotharingia was printed in three colors in the 1513 Strassburg edition of Ptolemy, and poly-color from engraved plates was attempted in the eighteenth century. See Woodward, et al., <u>Five Centuries of Map Printing</u>.

Chapter V

THE RACE TO SKIRT AMERICA

15. THE NEW WORLD

[untitled map of the Spanish Main].

Petrus Martyr d'Anghiera. Seville, 1511.
Map of the Caribbean from Martyr's *Legatio Babilonica Occeanea decas.*

Medium : woodcut.
Size of original : 190 x 280 mm.

PLATE : 10

When Columbus returned from his first Atlantic crossing in 1493, he visited the royal court in Barcelona to relate news of his success. One of the people present was Peter Martyr, a bright and well-educated man of northern Italian birth who was privy to all the latest news in cosmological doings. Later interpreting the boasts of Columbus, Martyr perceptively told a correspondent that the land from which the Genoese explorer had returned was the "western antipodes," dismissing from the outset Columbus' claim of having reached the Orient :

"A certain Colonus has sailed to the western antipodes, even to the Indian coast, as he believes. He has discovered many islands which are thought to be those of which mention is made by cosmographers, beyond the eastern ocean and adjacent to India. I do not wholly deny this, although the size of the globe seems to suggest otherwise, for there are not wanting those who think the Indian coast to be a short distance from the end of Spain." [142]

It was thus Martyr who is first known to have identified the concept of a "western hemisphere." [143]

Avidly interested in the new explorations, Martyr befriended Columbus, Vespucci, Cabot, Magellan, da Gama, and Cortés, among others. He was a member of the Spanish Crown's *Casa de las Indias,* the elite organization which guarded secret exploratory information; he was also priest and confident to Queen Isabella, and tutor to Ferdinand and Isabella's children. That he was a highly regarded observer is colorfully illustrated by the fact that Pope Leo X is said to have had Martyr's letters read to him at dinner. Leo X, often considered incompetent as a pope, was a dilettante of letters and art, and of some note in exploratory matters. In 1514 he issued the bull *Praecelsae devotionis,* which extended the *raya,* the line of demarcation dividing the New World between Spain and Portugal, around the other side of the earth through the Pacific Ocean. This had become necessary during his tenure in order to arbitrate recent claims to Asian lands which had of late muddled the existing 1494 settlement.[144] Martyr's letters had helped him keep abreast of such matters.

---

142. As translated in Parry, The Discovery of South America, except that this author has corrected that printing's "I do now wholly" to "I do not wholly."

143. Except for the theoretical quadrants described by Macrobius. See entry 3.

144. When the "raya," the line of demarcation, had been drawn by the 1494 Treaty of Tordesillas 370 leagues west of the Cape Verde Islands, neither eastern Asia nor the Australasian "spice" islands had been reached and thus were not yet a practical issue. Spain felt that Leo X's solution to the problem in 1514 favored the Portuguese, particularly after Magellan discovered the Philippines under the Castilian flag, and so the two countries

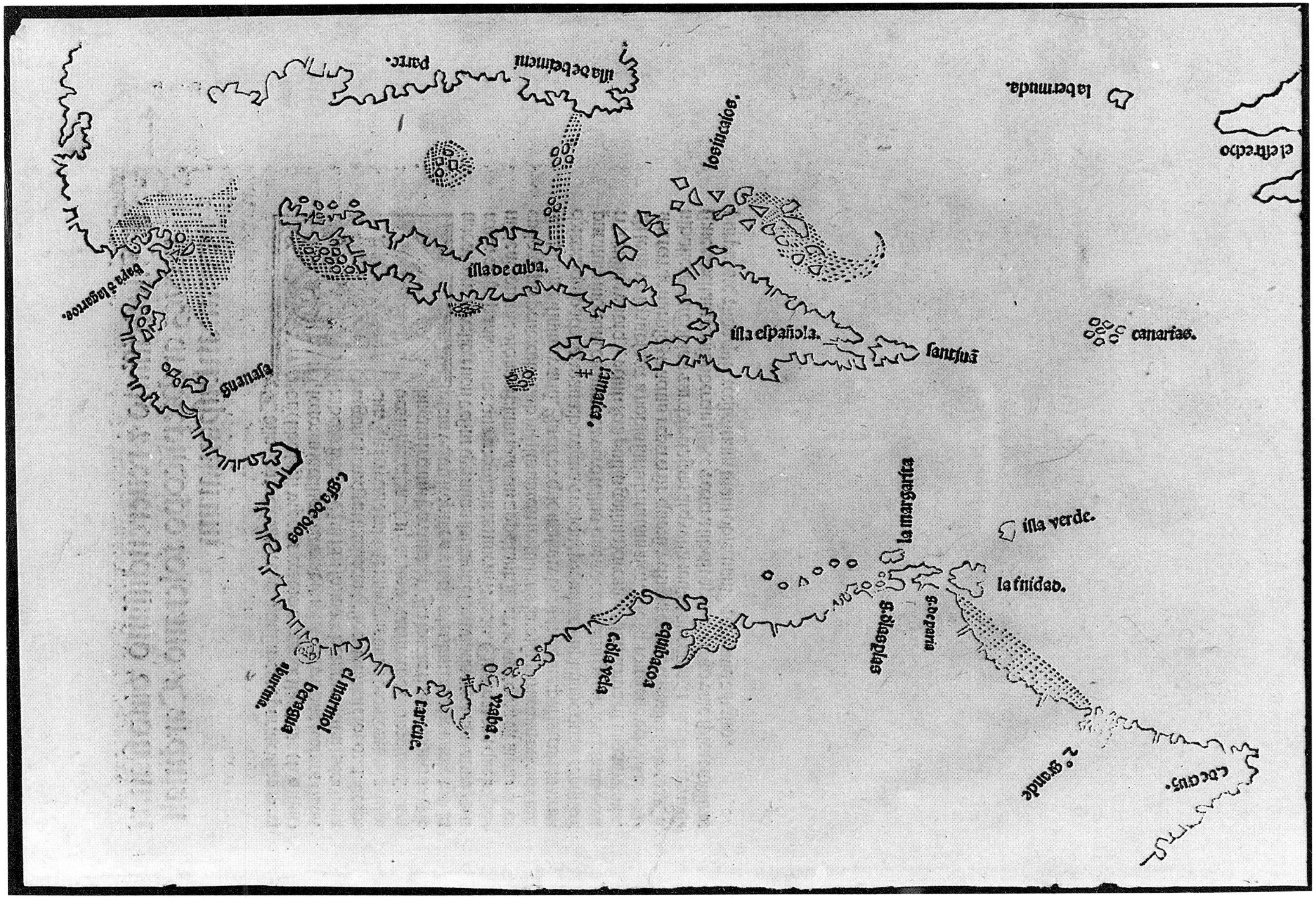

PLATE 10   Peter Martyr (d'Anghiera), Seville, 1511   (entry 15)

Several years into the exploration of the New World and after Columbus' death, Martyr set about writing a general history of the discovery of the New World. The book itself (though perhaps not yet the map) is known to have already appeared early in 1511, as on January 6 of that year Queen Juana ("la loca"; Isabella had died in 1504) made reference to Martyr's writings, *"especially a book containing the things which relate to the oceanic sea and its islands recently discovered."*

But revealing his country's exploratory secrets via a printed map was a brazen and unprecedented indulgence in Spain. The Crown was not eager to share the hard-won secrets which his map exposed, and in this same year of 1511 King Ferdinand issued a decree forbidding giving maps to foreigners. Although Martyr put on the verso of the map a beautifully printed epistle addressed to Cardinal Ximenex, archbishop of Toledo, arguing the relevance of a map to illustrate his book, the map is normally wanting from the already rare *Legatio*, suggesting that it was suppressed by the Crown.[145] Considering the political climate, it remains an enigma as to why a person of Martyr's standing would have risked making such a map at all. In part, Martyr's epistle states that:

> *"In order that the readers may more easily reach the recesses of our Nereids*[146] *(I mean our small books, such as the Oceanic Decade),*[147] *I have decided to add to the end of that decade the map herein engraved. Thus will those who desire to have an idea of the said treatises, and retrace in the mind that navigation, find there the direction . . . The square island, near Hispaniola, on the east side, they call St. John . . . on the left, follow with me that meandered, diversified, and infinite traject. It is the one of which so much has been said, where are the Dragon's Mouth, Paria, Curiana, Cauchieta, Cuquibacoa, Uraba, Veragua, and the other very large provinces which are said to be part of the Indian Continent. The land which is first seen to the west of Hispaniola, surrounded on all sides with islands (like a hen with her chickens), is Cuba, the large isle . . . At the north, marvellous lands and marvellous countries have been found, of which you can see the engraved representation on the right. I have omitted the Cannibal islands, the archipelago and others which are in the centre of them, so as to avoid confusion in this map."* [148]

Martyr charts the Caribbean Islands, part of South America, and possibly Florida (although no European encounter with Florida is known prior to 1513). A clue to the way Martyr's contemporaries perceived these lands fitting into the world puzzle can be noted in the above epistle, where he comments that the long "meandering" coastline (meso-/South America) is "said to be part of the Asian *('Indian')* Continent." Thus even for the *cognoscienti* of the Spanish court, the second decade of the post-Columbian era began without resolving the question of whether America was, in Martyr's own words, the "western antipodes," or rather a part of the Orient.

The mainland promontory west of Cuba bearing the name *baya d'lagartos* is sometimes believed to be the Yucatan, but is more likely the coast of northern Honduras, based on

---

reached their own compromise, without papal intervention, with the Treaty of Sargossa in 1529.

145. Additionally, the map may not have been ready for the first issue of the book, although it is hard to dismiss the element of government control. The next map printed in Spain was the woodcut in Pedro de Medina's Arte de Navigar (1545), which, though a fine map for its day, did not reveal enough data to present the "intelligence leak" that the Martyr map did, and so may have been officially tolerated. No "official" map was published in Spain until 1790. The earliest maps printed in Spain were a simplified Ptolemaic world map (Salamanca) and a map of Palestine (Zaragoza), both printed in 1498.

146. Literally a sea-nymph, daughter of Nereus.

147. Martyr's history was written in a seris of "Decades," the first (from which this map comes) being Occeanea decas.

148. Translation extracted from Harrisse, Discovery of North America.

Columbus' fourth voyage (1502-03).[149] No European knowledge of the Yucatan is recorded prior to 1517.[150]

Martyr's Florida, if it is indeed Florida, appears as *isla de beimeni parte* ("part of the island of Bimini"). Sometime before 1511 Indians in the Bahamas are said to have told their Spanish visitors that an island, named Bimini, lay to the north which had a spring which restored youth. Martyr is the first to have mentioned this island and its seductive "fountain of youth," which quickly became fashionable conversation in royal circles. In 1512 Ponce de León secured the right to discover and settle Bimini, resulting in his recorded "discovery" of Florida in 1513.[151]

Whether or not the Bimini that so entranced the Spanish court was in fact the Florida peninsula, based on earlier encounters, remains unknown. Yet Martyr's statement in his epistle that those "marvellous countries [at the north] *have been found*" implies that he, at least, believed that his Bimini was not based on fancy.

The island of Bermuda, discovered by Juan Bermudez in 1505, appears here for the first time.[152] Bermuda, the Canary Islands, and Trinidad all appear on close longitudes, an error markedly inconsistent with the uncanny accuracy of the map overall. This reveals either a profound ignorance on the part of Martyr as to the size of the Atlantic, or, more likely, that he conceived the map in a loose "isolario" style and did not intend to maintain scale across the ocean. To have done so would have required that he make all the map's features considerably smaller. As one so intimately in touch with current voyages to the Spanish Main, and one privileged to consult coveted government charts, it is unlikely that he accepted the diminutive proportions of the globe that had been proposed by Toscanelli and Columbus late in the previous century.

- - - - - - - - - - - - - -

149. See David Tilton, "Yucatan on the Peter Martyr Map?," Terrae Incognitae, Vol. XXI, p. 17-25 (1989).

150. In 1517 Francisco de Córdoba was the first known to have brought news of the Yucatan to Europe. It is believed, however, that two Spaniards, Gonzalo de Guerrero and Gerónimo de Aguilar, were shipwrecked on the shores of the Yucatan in 1511. Guerrero joined the Maya, but Aguilar was rescued by Cortés in 1519. Aguilar had learned the Mayan language and became Cortés' interpreter. A "true" depiction of the Yucatan in 1511 would of course mean a previous and unrecorded encounter.

151. The legend of a fountain of youth was an old one in Europe, and the Spanish may have resurrected it from a simple Indian report of, for example, limestone springs, found on the continent. Ponce de León, who had sailed with Columbus on his second voyage and became the conquering ruler of Puerto Rico, sailed north in March of 1513 in search of Bimini and its Fountain of Youth, resulting in the earliest recorded landfall in Florida and discovery of the Gulf Stream. Whether a fountain of youth was truly his objective is not known. Las Casas (who knew him) wrote that the taking of slaves was his real motivation.

152. Confusion has sometimes been occasioned by the omission of Bermudez' obscure first voyage; some bibliographers have been aware only of his later voyage of 1515, thereby leaving Martyr's Bermuda a mystery.

16. THE NEW WORLD

Tabula Terre Nove.

["Map of the new lands"]. Martin Waldseemüller, Strassburg, 1513. Map of the Atlantic coast of the New World from Waldseemüller's edition of Ptolemy's *Geographia*.

Medium : woodcut, with original hand color.
Size of original : 380 x 490 mm.

COLOR PLATE : IX

This is the first printed map specifically devoted to the New World to appear in an atlas. Unlike the map of Peter Martyr (previous entry), it charts a continental Atlantic seaboard, showing a continuous coastline stretching from 35° south latitude, where the mouth of the Rio de la Plata lies, to the latitude of the St. Lawrence River in North America. The map forms what appears to be a complete Gulf of Mexico and Atlantic Seaboard. Cuba is named *Isabella* after Queen Isabella of Spain, who more than King Ferdinand, was responsible for the Crown's acceptance of Columbus' proposal early in 1492.

When published in 1513, six years had elapsed since Waldseemüller had coined the name *America* to denote South America. But here he abandons that term, instead identifying South America as having been discovered by a certain Columbus sent by the King of Spain :

> *"Hec terra cum adiacentib insulis inuenta est per Columbus ianuensem ex mandato Regis Castelle"* ("This land with its adjacent islands was discovered by Columbus, sent by the king of Castile").

In addition, Columbus' *Parias* (in Venezuela) is no longer displaced to North America in an apparent attempt to separate his discoveries from those of Vespucci, as Waldseemüller had done on his 1507 work.[153] Both these factors appear to be an admission by Waldseemüller that he had been misled by the accounts of Vespucci's adventures, and that he regretted christening the continent with his name.

Waldseemüller's northern coastline is more mysterious, and presents two essential questions: whether it is autonomous or connected to Asia, and whether its geographic features are those of Asia or America. The more common viewpoint is that it is not connected to Asia and that the protruding land just northwest of Cuba is "true" Florida, the *Bimini* of Peter Martyr. If so, this would verify European exploration of the peninsula prior to Ponce de León, and would also mean, remarkably for the day, that South, Central, and North America were recognized as sharing a continuous coastline as components of a single mammoth continent. But this coastline may not represent North America at all.

The counter argument is that the northern coastline is in fact the Orient, into which some North American data has been incorporated, and that our "Florida" may have been perceived as a feature of a Southeast Asian landscape. Although Waldseemüller had earlier depicted North America as a new land distinct from Asia, in 1516 he produced another map of the world, the geography of which more closely matches the present work, in which he reversed that opinion.[154] On that map our enigmatic "North American" mainland is labelled

---

153. This interpretation is supported by the fact that Lorenz Fries, who perpetuated use of the term "America" on world maps he cut in 1520 and 1522, re-cut the present map in 1522 with only cosmetic alteration <u>except</u> that he revives Columbus' "Parias" for North America, using it to designate all of the mainland above the Caribbean. See page 52 in entry 13 (Waldseemüller/Apianus), and footnote 128.

154. For Waldseemüller's depiction of America and Asia as independent continents see entries 11 (Waldseemüller) and 13 (Apianus). On his "Carta Marina" of 1516 he depicted them as a single contiguous continent, though he evades the nature of their connection by omitting 108 degrees of longitude.

Terra de Cuba Asie Partis ("Land of Cuba, part of Asia"), bowing both to Columbus' denial of an insular Cuba and to his early insistence that he had reached the Orient, and demonstrating that our *Tabula Terre Nove* shows the new discoveries as previously unknown shores of the Orient, rather than as a new *world*.

Cuba itself ties in with this interpretation. Waldseemüller's orientation of the island suggests that it had earlier been connected to his "Florida," supporting the view that the peninsula, even if not a "true" Asian feature, is in any case the vestige of Columbus' peninsular (Asian) Cuba, not Ponce de León's Florida. About the time Waldseemüller was preparing our map, Martyr was still allowing this Columbian view of Cuba the benefit of the doubt :

> "*the Ilande of Iohanna or Cuba . . . westwarde, it is the beginninge of India beyond the ryver Ganges: And Eastewarde, the furthest ende of the same: which thinge is not contrary to reason for as muche as the Cosmographers have lefte the lymites of India beyonde Ganges undetermyned.*" [155]

But correlation of the two maps is inconclusive. On the 1516 map, Waldseemüller leaves the continuity of the Atlantic coastline ambiguous, while here he shows a virtually continuous coastline down to South America, which was from the outset recognized as a land separate from Asia on maps of this genre.[156] Both the 1516 map and the present 1513 *Terra Nove* share a somewhat common ancestry with the "Cantino" Planisphere (manuscript, 1502), which initiated the "Florida" pattern adopted by this and other maps, but is ambiguous as regards the connection of new discoveries to Asia.[157]

- - - - - - - - - - - - - -

17. MEXICO CITY AND THE GULF OF MEXICO

[untitled].

Hernan Cortés, Nuremberg, 1524. From Freidrich Peypus, *Praeclara Ferdina[n]di Cortesii de Nova Maris Ocenai Hyspania Narratio . . .*

Medium : woodcut.
Size of original : 310 x 465 mm.

PLATE : 11

Hernan Cortés was a boy of seven when Columbus first reached American waters. He took part in the conquest of Cuba in 1511, and from there in 1519 he sailed west to the mainland, destroying the Aztec capital of Tenochtitlan (*Temixtitan*, Mexico City) in 1521. By 1522, a mere three decades after Columbus' maiden landfall, he had conquered all of Mexico for Spain.

This map accompanied the second "letter" of Cortés, which was written in 1520 and published four years later. Its right section shows an enlarged plan of *Temixtitan,* then probably the most sophisticated community in the hemisphere. This is the first printed plan

---

155. Martyr/Eden, Decades, p. 13 (originally published 1511).

156. A few specialized maps did show South America as part of Asia, but this does not belong to their tradition. They include the Juan de la Cosa manuscript map of 1500 (but probably circa 1508), and the Finaeus maps of 1531 and 1534 (entries 19 and 20).

157. See footnote 135 for Cantino.

of an American city. The metropolis, which was the pivot of Montezuma's empire, sits within Lake Texcoco and is connected to the outer shores by causeways. According to Gómara,

> *"this Citie is built upon the water, even in the same order as Venice is. All the body of the Citie standeth in a greate large lake of water."*[158]

Three types of streets were described, one of earth, one of water, and one which was half firm ground and half canal.

The fact that the city is supplied with water from neighboring mountains via an aqueduct (visible at the top of the map with the legend *Ex isto fluvio conducut Aqua in Civitatem*) reveals that the lake water was not potable. Gómara explains that

> *"although this Citie is founded upon water, yet the same water is not good to drynke, wherefore there is broughte by conduit water from a place called Chapultepec, three myles distant from the Citie . . . the water is brought thence in two pypes or Canalls in greate quantitie, and when the one is foule, then all the water is convayed into the other, til the first be made cleane."*

At the center is the temple of Teocalli, which is made of stone and *"is square, & doth containe very way as much ground as a crossebow can reach levell."*

The city appears well-ordered, with parks, Montezuma's palace (upper left), and Mexicans shuttling between city and mainland in canoes. By the time this map was published in 1524, *Temixtitan* had been levelled by Cortés, and Spanish Mexico City had been founded in its place.

The left portion of the map shows the entire Gulf of Mexico. Cortés states in his letter that Montezuma had given him a map of the coast, apparently influencing the present work and, if so, making it the first published map to be based partly on Amerindian sources.[159]

A Spanish pilot by the name of Antón de Alaminos also figured strongly in the events leading up to Cortés' adventures and this map. Alaminos, whose introduction to the American mainland dated back to 1502 when he was with Columbus in Honduras, piloted Ponce de León to Florida and was largely responsible for the discovery of the Gulf Stream which made the return trip to Spain from her colonies practical. In 1517 he guided Francisco de Córdoba west from Cuba, where they reached the Yucatan after floundering in a bad storm.[160] As a result of this circumstance by which the Yucatan was "discovered," Cortés' map shows it as an island; the misconception prevailed throughout the first half of the century.[161]

The following year (1518) Alaminos piloted another voyage from Cuba, now under the command of Juan de Grijalva. This expedition brought back news of a land of great splendor lying to the west, and it was as a result of this that governor Velázquez of Cuba sent Cortés west in 1519. Upon reaching Vera Cruz, however, Cortés' people disassociated themselves from Velázquez, declared their direct allegiance to Spain, and pronounced Cortés their leader. From there began his famous conquest, one of the most ambitious and ruthless rampages of the era.

In this same year of 1519, governor Garay of Jamaica sent Alvarez Pineda to discover a sea route through the mainland to the Orient. The search for such a route in meso-America had intensified since Balboa's sighting of the Pacific from Darien in 1513. With Alaminos as pilot, they skirted the entire Gulf Coast through Florida, losing many men to disease and fighting. The survivors eventually reached Cortés at Pánuco in Vera Cruz, shown as *Provincia Amichel,* the name Garay had selected for a proposed colony.

---

158. Gómara/Nicholas, The Conquest of the Weast India, p. 192-93.

159. Except possibly for Peter Martyr's Bimini. See entry 15.

160. See footnote 150.

161. The first printed map to show the Yucatan correctly as a peninsula of the Central American isthmus was the cordiform world map of Gemma Frisius (1544), which appeared in his revision of the Cosmographia of Apianus (originally published 1524).

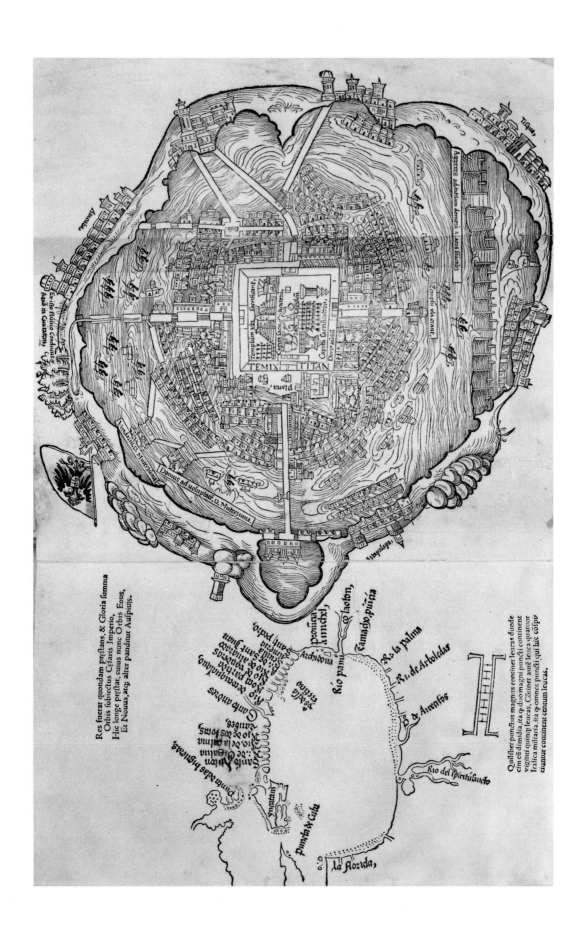

PLATE 11   Hernán Cortés, Nuremberg, 1524   (entry 17)

Cortés drew heavily on Pineda's experiences. Pineda had conclusively shown that the Gulf formed an unbroken coastline through Pánuco, thus fitting together the final connecting piece of the American continental puzzle. The mainland was continuous from Florida through South America, and so Spain, in her search for a passage through the New World to the Orient, turned her eyes to the little-known shores north of Florida.[162]

On the north (bottom) the *Rio del spiritu sancto* appears here for the first time on a printed map.[163] This river "of the holy spirit" is often thought to be the Mississippi, although its identification as such is far from clear. Based on its described shape, and on the report that it had a large village at its mouth and forty more within six leagues up river, it more likely represented Mobile Bay and the Mobile River. *Rio del spiritu sancto* continued to appear on maps through the end of the seventeenth century, by which time it may in fact have "become" the Mississippi, neatly transposed when later geographers assumed it to be such.

Though a Spanish work, Cortés' map was published in Germany because the Crown's paramount stakes in geographic data did not permit the dissemination of such sensitive information.

- - - - - - - - - - - - - -

18. NORTH AMERICA

[untitled].

Benedetto Bordone, Venice, 1528. From Bordone's *Isolario.*

Medium : woodcut, with original hand color.
Size of original : 85 x 150 mm.

COLOR PLATE : X

One type of navigational aid commonly used by pilots sailing Mediterranean waters was the *Isolario,* a book containing descriptions and maps of the sea's islands. The first printed Isolario appeared in 1485; that work, by Bartolomeo dalli Sonetti, was limited to the Aegean islands. The second printed Isolario was Bordone's work of 1528, which attempted to chart the islands of the entire world. North America, vastly undersized and severed from South America by a strait, figured logically into Bordone's compilation of the world's islands. This is the first printed map specifically of North America.

Bordone's "island" of North America bears the single label of *Terra de lavoratore* (i.e., Labrador). The term comes from *el lavrador* ("the farmer"), nickname for a Portuguese-Azorean adventurer by the name of João Fernandes.[164] Fernandes may have tried his luck at

---

162. About the time the Spanish Crown learned of Pineda's and Cortés' failure to find a route through meso-America, she also learned from a defector of Magellan's voyage that that expedition had shown South America to extend unbroken further south than expected. See entry 27 (Bellero), page 92.

163. The Rio del Spirito Santo is found on the Garay manuscript map of 1519 and the "Turin" manuscript map of circa 1523.

164. Such a nick-name did not necessarily imply ordinary social-standing or poor background.

western voyages under the Portuguese flag as early as Columbus had under the Spanish flag.[165] At the turn of the century, however, Fernandes emigrated to England and quickly secured letters patent from Henry VII for the discovery and possession of new lands. Two theories might explain his sudden change of allegiance: in Portugal he may have been out-ranked by the more lavish Corte-Real brothers in securing rights to any lands discovered, or rather Fernandes might simply have been enticed by a better offer from Bristol merchants.[166] Sailing in 1501 under his new privilege from Henry VII, he is thought to have discovered Labrador, with use of his nickname to denote the region dating as early as 1502.[167] His true landfall has long been disputed, however, and some historians argue that he reached Greenland rather than any part of America.[168] A cavalier attitude on the part of some early mapmakers has only compounded the issue, as *Labrador* is sometimes found as a wedge-shaped trace of land above America, later being assimilated into the American continent itself. Bordone's map offers nothing to the unresolved issue of what shores Fernandes reached.

That Bordone's little map of *Terra de lavoratore* is the North American "continent," not self-evident from its geography, is demonstrated by two points. The simpler is by its context in Bordone's world map in the same *isolario*. Perhaps even more convincing is the term *stretto pte del mondo novo* found at the southern extreme of our *Terra de lavoratore,* the "new world" with which it shares the strait clearly referring, at this time, only to South America (and as South America is in fact designated on Bordone's world map). However, although the landmass is *functioning* as an autonomous North America, its geography is rooted in a primitive depiction showing North America as eastern shores of an elongated eastern Asian coast. Bordone has taken such an earlier map, extracted its "Amerasian" section, and added an arbitrary western coastline to complete it. The two Rosselli world maps of circa 1508 are likely candidates. Geographically, they both show an ancestral connection to Bordone's, and in fact Bordone copied the projection of Rosselli's oval map for his world map. And Rosselli, like Bordone, designated all of the North American discoveries after João Fernandes.[169]

Little, if any, true North American geography has been incorporated into Bordone's originally Asian landscape.[170] The map is, for example, peculiarly void of any trace of the Florida (or pseudo-Florida) peninsula already found on other printed maps for two decades. In any case, the narrow section of North America above the *Stretto pte del modo novo*

---

165. Records in Lisbon and the Azores show that he sailed in voyages of discovery under license from King John in 1492-95. Some authorities (e.g., Samuel Morison) do not believe this, and in any case no discoveries are recorded as having been made. In 1499 he received license from Manuel I for further (?) exploration, although as he transfered to England shortly afterwards that license appears to have never been used.

166. Gaspar Corte-Real had received a license for the discovery of new lands from Manuel I in 1500, which may have conflicted with any similar request by Fernandes (see page 56). In Bristol, Fernandes joined a group of merchants which had links with the Azores; this group included two fellow Azorians.

167. It is found on the "Cantino" portolan chart of the world made for the Duke of Ferrara in 1502. See footnote 135.

168. This contributed to the mis-identification of the present Bordone map as other than North America; e.g., Bagrow (History of Cartography, p. 64) refers to this work as a map of Greenland.

169. There are two Rosselli world maps from circa 1508, one on an oval (which projection Bordone copied for his world map) and a sea chart, both untitled. The sea chart clearly records "Tyerra de labrdor," and the oval map in all probability shares the same place-name, but it is essentially illegible.

170. There is however a possible, paramount exception to this. Just to the left of the word "Terra" is a large bay and flat east-west coast which closely resembles the depiction of New York Bay and southern coast of Long Island following Verrazano which is found on the 1548 "Tierra Nueva" of Gastaldi and the 1556 "La Nuova Francia" of Ramusio. Although Bordone's use of Verrazanian data by 1528 is conjectural and in fact quite unlikely, the similarity of these features to those on the Gastaldi/Ramusio maps is striking, if coincidental.

corresponds roughly to southern North America (lying at 35° north latitude), and the *stretto* itself is approximately in the region of Mexico where Cortés and Garay had hoped to find one, and indeed where some earlier maps timidly alluded to one.[171] The lower border of the map is drawn along the Tropic of Cancer.

In the Atlantic, Bordone has abandoned scale to allow him to include Fernandes' native Azores, and the fabulous island of Brasil from old Irish legend. The other island shown by Bordone, *Asmaide,* has not left such a rich history as Brasil Island, but it is also found near Brasil on other maps, such as the 1513 *Terre Nove* of Waldseemüller (entry 16).

- - - - - - - - - - - - - -

19. THE WORLD

Nova, et Integra Universi Orbis Descriptio.

["New description of the entire world]. Orontius Finaeus, Paris, 1531 (1540). Map of the world on a double-cordiform projection.

Medium : woodcut.
Size of original : 290 x 420 mm.

PLATE : 12

One of the most compelling questions facing mapmakers in the first half of the sixteenth century was that of the relationship of the New World to the Orient. The question almost always pertained to North America; South America had grown on its own as truly a new world.[172] If South America bore any continuity with the Orient, it was only via Central and North America.[173]

But with this map, Finaeus adopts a different and very radical stance : North and meso-/South America are each individual extensions of Asia. This is the completion of a cycle that began late the previous century with the demise of the old landbridge connecting Southeast Asia to Africa on Ptolemaic maps. After the Portuguese succeeded in reaching the Indian Ocean by way of Africa, some cartographers opened the landbridge but left its residue in the form of a very large "extra" Southeast Asian peninsula.[174] By coincidence, the peninsula occupied the approximate place that meso-America would if the West Indies were Australasian islands, and thus its appearance on some maps lent credence to the belief that meso-America was part of an Asian peninsula, the *Cattigara* of Ptolemy. Here in the final phase, Finaeus has predicated this map on this very contention, transforming the earlier vestige of Ptolemy's Africa-Asia landbridge into "true" meso- and South America. Thus there are three Asian peninsulas in Finaeus' Indian Ocean (upper left-side of map). Counter-clockwise they are India, the Malay peninsula, and meso-/South America, the last being mapped as a descendent of Ptolemy's landbridge, complete with the metropolis of *Cattigara*. The dominating new forces which consummated this evolution were the expeditions of Vasco Nuñez de Balboa, who traversed the Central American isthmus in 1513, and Ferdinand Magellan, word of whose circumnavigation of the earth (1519-22) spread through Europe during the later 1520s.

---

171. E.g., see entry 16 (Waldseemüller).

172. The portolan chart of Juan de la Cosa is ambiguous but implies that both North and South America are parts of Asia (1500, but probably circa 1508).

173. E.g., the 1548 "Carta Marina" of Gastaldi (entry 24).

174. E.g., the world maps of Waldseemüller (entry 11), Apianus (entry 13), anonymous/Holbein (entry 21).

PLATE 12

Orontius Finaeus
Paris, 1531

(entry 19)

Magellan was a Portuguese seaman sailing in the employ of Spain. Of the five ships and 270 people who departed Lisbon in 1519, one ship and thirty-one men returned (eighteen with the expedition itself, and thirteen separately). Magellan himself perished in the Philippines. The voyage clearly was one of the most staggering achievements of navigation, dwarfing that of Columbus barely three decades earlier. But Finaeus' map, while strongly influenced by the voyage, equally demonstates how poorly it was understood. The voyage appears to have reinforced, rather than dispelled, the belief that Ptolemy's *Cattigara* was in reality part of the New World. According to Martyr, Magellan and his crew had *Cattigara* on their minds,

> *"intendying as much as were possible, to approche to the cape cauled of the owlde wryters Cattigara: The whiche is not founde as the owlde Cosmographers have discribed it, but is toward the north abowt xii Degrees as they afterwarde understode."* [175]

Thus Finaeus has accomodated the belief that Central America was an Asian peninsula, and the place-names of the Orient cohabit with those of America.[176] North America is shown as a continental extension of Asia, with only an "Amerasian" gulf to partition its soil from that of Asia proper. Magellan's extraordinary voyage, rather than demonstrating that America and Asia were distinct continents, has here cultivated the opposite conclusion. As explained by the theorist Johann Schöner, whose globe of 1515 may have influenced Magellan,

> *"thanks to the very recent navigations . . . by Magellan . . . it has been ascertained that the said country* [America] *was the continent of Upper India, which is a part of Asia."* [177]

In order to adopt this configuration Finaeus had to contort his geography in two ways. First, he had to place nomenclature from the east coast of Asia on the west, rather than east, coast of "Amerasia." This was the only way that he could map meso-/South America as the "Golden Peninsula" *(Aurea Chersonesus)* of Ptolemy while not contradicting the knowledge that Marco Polo had sailed from Zaiton to the Persian Gulf on his return to Venice from China. Secondly, Finaeus necessarily minimized the extent of ocean between America and Asia. As with *Cattigara* and the depiction of South America as an Asian sub-continent, this is ironic : Magellan's maiden circumnavigation of the earth should have instead painfully impressed upon Finaeus the vast extent of ocean that lay between the two continents.

Magellan's strait, or series of straits, separating the southern tip of continental South America from the islands of Tierra del Fuego is recorded here nine years after the return of the survivors of his expedition. Finaeus labels the waters just west of it in his honor: *Mare magellanicum.*[178] This is the first verbal reference to Magellan's voyage on a printed map.[179]

With regard to Europe's perception of Magellan's circumnavigation, it is interesting to note that as late as 1538 the Spanish cosmographer Pedro de Medina, attempting to

---

175. Martyr/Eden Decades, p. 222.

176. One theory holds that Ptolemy's Africa-Asia landbridge did in fact represent America. See entry 5 (the "Rome" Ptolemaic world).

177. Schöner's globe shows a strait between America and a southern continent at about 45 degrees south latitude. Extract from his Geographical Tract of 1533, and quoted from Polk, The Island of California.

178. Magellan's term "Pacific" did not appear until nine years later ; see page 84 in entry 23 (Münster).

179. Although reference to Magellan is found, e.g., on the Pedro Reinal (anonymous) map of circa 1522 and the Ribeiro (anonymous) map of 1525; see Portugaliae Monumenta Cartographica, Lisbon, 1960. Also, influence from the voyage is found on the printed maps of Franciscus Monachus (circa 1527) and Robert Thorne (1527?).

demonstrate the sphericity of the earth, offered various circumstantial arguments but never invoked Magellan's voyage as irrefutable proof.[180]

Magellan's strait posed a new question for mapmakers: What was the nature of the land which formed its southern shores? Most geographers interpreted it to be continental in nature, thus creating out of Magellan's *Tierra del Fuego* the *Terra Australis,* or antipodal southern continent long thought to exist as a balance to the known northern landmasses. Finaeus has here christened the trend, Gastaldi and the Italian school soon adopted it, and the Dutch/Belgiun school, notably Mercator and Ortelius, would continue it. The opinion was not, however, unanimous ; Sebastian Münster, for example, believed that

> *"the land which* [Magellan] *had on his right hande, he doubted not to be mayne lande: and that on the left hand, he supposed to bee Ilandes."* [181]

In any case, Magellan presumably believed that *Tierra del Fuego* was at least large, for had he suspected otherwise he logically would have tried to sail around it rather than navigating its tricky straits.[182] The fact that *Tierra del Fuego* was only a series of islands, with open ocean to the south, remained unknown for nearly a century.[183]

Finaeus' polar oriented double cordiform projection allowed him to confront *Terra Australis* directly. His depiction bears a noticeable resemblance to true Antarctica in general contour, and as a result the map has been studied as evidence of lost ancient knowledge.[184] "True" Antarctica is not known to have even been sighted until the early nineteenth century.[185] Although it can be assumed that many monumental Renaissance and ancient voyages of discovery remain unknown to us, history as we accept it clearly precludes the possibility for unrecorded voyages by Finaeus' time to have actually charted Antarctica so completely. The origin of his *Terra Australis,* if not a fluke of whimsy, remains a mystery.

Along the Indian Ocean shores of *Terra Australis* lies the Brazil of South America *(Brasielie Regio).* The transposition of Brazil to a hypothetical southern continent is found as early as the 1515 globe of Schöner, and appears to have arisen from at least two separate sources. Schöner seems to have displaced it in an attempt to follow an account of a Portuguese voyage along the Brazilian coast. Other proponents of an antipodean Brazil appear to have read a confusing account of Cabral's 1500-01 voyage by the Venetian ambassador Il Cretico.

Insight into Il Cretico's influence may be had from Mercator, who places a "parrot kingdom" (i.e., the Brazil of Cabral) in the southern continent on his 1538 world map (which used the present Finaeus map as a model), as well as on his globe of 1541 and world map of 1569 (see entry 31). Mercator sheds light on the matter by explaining, via a legend on the 1569 map, that when the Portuguese discovered the *"Psitacorum regio"* ("parrot kingdom") while en route to *Callicutium* (India), the winds which had blown them off course were the *libeccio* (southwest winds); as they had followed the land's coast for two thousand miles without finding its end, they therefore must have reached the southern continent. Mercator

---

180. Pedro de Medina, <u>Libro de Cosmographia</u>, question 72.

181. Münster depicts the Tierra del Fuego as a very large single island on his world map of 1540; this same model can be seen on Gastaldi's "Carta Marina" of 1548 (entry 24, plate 17). Quote from Munster/Eden, <u>A Treatyse of the Newe India</u>.

182. In truth, of course, the voyage around Cape Horn can be more treacherous than the Magellan Strait.

183. The insularity of Tierra del Fuego was suspected by Drake (1577-80) and proven by Le Maire (1615-17).

184. E.g., Charles Hapgood's rather far-fetched study in <u>Maps of the Ancient Sea Kings</u>.

185. James Cook crossed the Antarctic Circle in 1775 on his second voyage, reached the South Georgia and South Sandwich Islands, and came within fifty kilometers of the continent but could not see it because of fog. In 1819 William Smith, also British, discovered the South Shetland Islands, and finally, the following year, he and James Bransfield approached part of the Antarctic Peninsula. A sealing ship from Connecticut under Nathaniel Palmer reached the continent later the same year.

placed the explanation in his *Terra Australis* due south of southern Africa, indicating that he believed Cabral was already rounding the Cape of Good Hope, rather than the corner of West Africa, when the *libeccio* forced them to the parrot land.

Finaeus displaces it even further. His placement suggests that he envisioned them already sailing north towards India when the detour occured, more consistent with the German pamphlet *Copia der Newen Zeytung ausz Presilig Landt* (i.e., "News from Brazil"), which is believed to have guided Schöner. That publication reported that a voyage sailed south of the "Cape of Good Hope," apparently meaning not the tip of Africa but rather an *allegorical* reference to the southern tip of America, its presumed counterpart.[186] The report was, however, confusing :

> *"They reached the Cape of Good Hope, which is a point extending into the ocean, very similar to Nort Assril, and one degree still further. When they had attained the altitude of the fortieth degree, they found Brazil, which had a point extending into the sea . . . After they had navigated for nearly sixty leagues to round the Cape, they again sighted the continent on the other side, and steered toward the north-west . . . Driven away by the Tramontane, or north wind, they retraced their course, and returned to the country of Brazil."* [187]

The term "Nort Assril" is unclear but has been taken to mean the (southern) point of Africa. But it appears that Finaeus took the Cape of Good Hope reference literally, and so charted his antipodean Brazil to the southeast of Africa. The voyage to which the newsletter refers probably reached the Rio de la Plata, and thus may have influenced Schöner's mapping of a strait between South America and a southern continent (the southern shores of the Plata being that continent), which in turn may have influenced Magellan.

Like his southern continent, Finaeus' depiction of Greenland is extraordinary, if perhaps less inexplicable. While many contemporary maps depict Greenland as a peninsula of Asia or Europe, Finaeus shows it correctly as an island and in remarkably accurate fashion. Even the island's westerly bulge above Baffin Bay is primitively represented, an amazing feature at this early date, if not simply cartographic luck. North of Greenland, Finaeus shows the Arctic region as four large islands following a concept apparently originating in a fourteenth century treatise already used by Ruysch in 1507.[188]

The imprint in lower center of Finaeus' map originally bore the date of 1531, at which time it was sold as a separate sheet. In 1532 it was published in the Paris issue of Johann Huttich's *Novus Orbis Regionum* with the 1531 date unchanged; the present issue bears the date 1540, and was published in Pomponius Mela's *De Orbis Situ* of that year.

- - - - - - - - - - - - - -

---

186. See page 104 in entry 31 (Mercator).

187. Quoted from Harrisse, The Discovery of North America, p. 486.

188. See pages 104-05 of entry 31 (Mercator) for reference to treatise; see entry 12 for Ruysch.

20. THE WORLD

Cosmographia universalis ab Orontio olim descriptio. Joannes Paulus Cimerlinus Veronesis in aes incidebat Anno 1566.

["Map of the world by Orontio {Finaeus}" . . .]. Giovanni Cimerlino, Venice, circa 1566 (after the woodcut map of Finaeus, Paris, 1534).

Medium : copperplate engraving.
Size of original : 520 x 580 mm.

PLATE : 13

The previous map of Orontius Finaeus (entry 19) was constructed on a projection which split America along its seams. While that double-cordiform rendering had permitted an ideal view of the changes to the world image wrought by Magellan's circumnavigation, it left the equatorial regions of the earth confusingly disjointed. Here, however, Finaeus presents the same geography on a "true" cordiform projection, a single heart-shape with no truncation. With this perspective, evidence of another incursion into the Pacific, that of Vasco Nuñez de Balboa, is more legible.

It was with Balboa's crossing of the meso-American isthmus in 1513 that America first yielded a clue to the nature of its western shores. Through the region marked *Dariena* on the map, Balboa

"passed over the daungerous mountaynes towarde the South sea [and] *learned by report that in the prospect of those coastes there laye an Ilande aboundynge with pearles of the greatest sorte."* [189]

Having successfully crossed Panama, he beheld the Pacific lying to the south from his vantagepoint in the mountains of Darien. The frustrating impenetrability of the New World had been finally breached. It is not clear, however, precisely what waters Balboa believed he was gazing at. Balboa may have envisioned the world as did, for example, Waldseemüller on his map of 1507 (entry 13), with a "true," if still diminuitive, "Pacific" Ocean. In this view Balboa would have correctly perceived that these were the waters described by Marco Polo and that he, rather than his predecessors in the Caribbean, had finally reached the China Sea from the east. Conversely, he could have assumed that the isthmus he had crossed was part of the landbridge which had classically rendered the Indian Ocean a closed sea, that he in effect had just pierced the right border of the world map as shown by Ptolemy, that his South Sea was Ptolemy's *Magnus Sinus,* and that *Cattigara* therefore lay somewhere along his new coast. That is how Finaeus interpreted the events.[190]

Finaeus believed that America and the Orient were partitioned only by a gulf, in effect showing North America as Asia *Extrum Cathay* in a way analogous to earlier conceptions showing eastern Asia as being India *Extrum Ganges.* North America is partitioned from China only by a gulf similar to that which separates India from Arabia or Southeast Asia from India, with the Spice Islands squeezed into this "Amerasian" gulf. The data brought by the survivors of the Magellan voyage had failed to impress upon Finaeus the true size of the Pacific Ocean.

As a result of this confusion, the recent landfalls of American explorations co-exist with the worlds and place-names lingering from Marco Polo and the medieval mind. Although the regions of *Florida, Francesca, Baccalear,* and *Los cortes* properly occupy the eastern seaboard

---

189. Martyr/Eden, <u>Decades</u>, verso of p. 139.

190. See pages 26-27, entry 8 (Waldseemüller/Ptolemy).

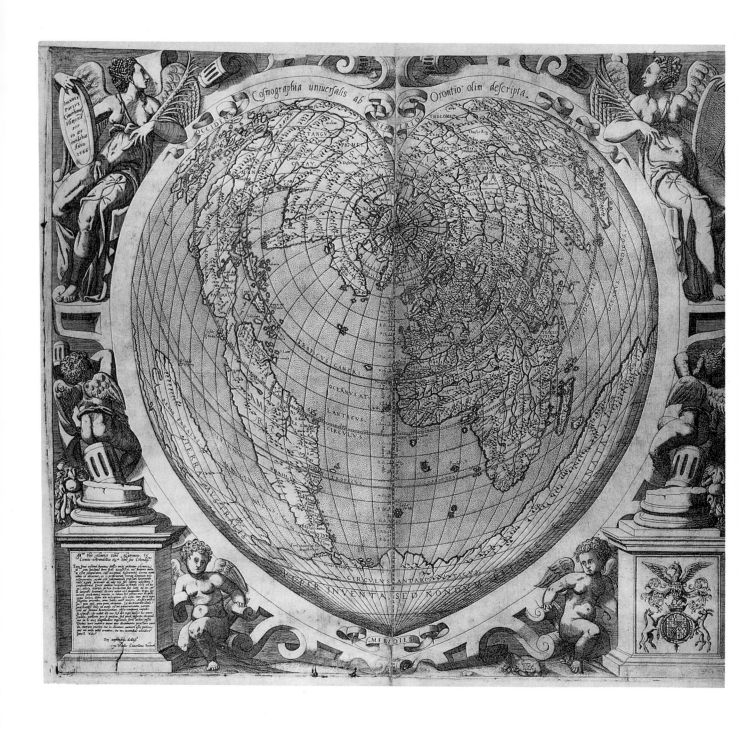

PLATE 13   Giovanni Cimerlino, Venice, 1566   (entry 20)

of North America, in the region of present-day Texas we find *Catay,* or Cathay (China).[191] West of Cathay, in the vicinity of what is now the American Southwest, Finaeus plots *Tangut,* the northwestern part of China where Marco Polo encountered the escorts of Kublai Khan. To the south, *Messigo* (Mexico) and *Temistita* (Mexico City) are simply regional names in the province of *Mangi* (Manzi, or southern China), whose splendor Polo described as being *"on such a stupendous scale that no one who hears of it without seeing it for himself can possibly credit it."* *Messigo* is in fact shown to the *west* of Cathay. Desert is shown stretching across the Great Lakes region. A realm of pygmies *(Pig Mei),* probably inherited from early Norse reports of *Skraelings* in Greenland, lies just east of the desert.[192] Off the western shores of the New World lies the Moluccan Sea *(Moluce Mare).*

The map, originally prepared in woodcut by Finaeus for Francis I in 1534, is here copied in copperplate by the Veronese engraver Cimerlino about 1566.[193] It is framed in an elaborate array of angels, cherubs, columns, and filigree, bearing a dedication to Henry Fitzalan, Earl of Arundel.

- - - - - - - - - - - - - -

21. THE WORLD

Typus Cosmographicus Universalis.

["A Map of the Whole World"]. [anonymous] (Hans Holbein? / Sebastian Münster? / Simeon Grynaeus?), Basle, 1532. Map of the world on an oval projection, from the Basle edition of the *Novus Orbis Regionum . . .* of Johann Huttich.

Medium : woodcut.
Size of original : 355 x 545 mm.

PLATE : 14

A few months before Johann Huttich's *Novus Orbis Regionum* was published in Paris with the world map of Finaeus (entry 19), it had already been published in Basle with this vastly different world map, an oval projection flanked by an elaborate border attributed to Hans Holbein the Younger. The anonymous map is from a geographic viewpoint a largely regressive work, though nonetheless an attempt to give order to diverse and confusing elements. Its most interesting geographic feature is its reconciliation of current data with Columbus' insistence that Cuba was part of continental Asia. To achieve this, the map's author has combined and modified features found on the two monumental world maps of Martin Waldseemüller, done in 1507 and 1516.

---

191. Although the term "Cathay" became synonymous with "China," it more correctly denoted only the region of the "Kara Khitai," a semi-nomadic people who ruled an independent state near the northwest border of China and whose military adventures helped to spark the myth of Prester John. Cathay proper, the country of the Kara Khitai, was conquered by Ghengis Khan in the early thirteenth century.

192. "Skraelings" were a race of small people reported by early northern voyages to Greenland. The Eskimos in Greenland may well have averaged a shorter physical stature than their Norse visitors. See Morison, The Northern Voyages, p. 53-56, and p. 420.

193. Harrisse, The Discovery of North America, considers the map's model to date from 1521, shortly before Fineaus was imprisoned. In any event, as the map records the discoveries of Magellan, in its present form its geographic content could not have been established until sometime after the return of that voyage's survivors in 1522.

The configuration for North America is extracted from the 1507 work, but it has been given the identity of Cuba as per the 1516 map. That map, in a posthumous concession to Columbus, identified North America as *Terra de Cuba Asie Partis* ("Land of Cuba, Part of Asia").

But by adopting Waldseemüller's 1507 rather than his 1516 geography, this 1532 map shows North America as a separate land not connected to Asia, and therefore identifies it only as *Terra de Cuba*, deleting the *Asie Partis*. The map's author accepted Columbus' claim that Cuba was not an island, accepted Waldseemüller's 1516 map in equating Cuba with North America, but did not accept Columbus' claim that it was part of Asia. The problem remained, however, as to what to do with the *island* of Cuba. It had been curiously retained, without identity, on the 1516 Waldseemüller map; the author of the present map logically deletes one of the two large islands representing Cuba and Hispaniola, but indecisively retains the names of both islands, *Isabella* (Cuba) and *Spagnola* appearing above and below the remaining island.

In the north, the landfall of the Corte-Reals is neatly contained in an island, *Terra Cortesia*. The mythical island of Antilia has now evolved into the "real" Antilles, *Insulae Antigliae*. Japan, still more than a decade pre-"discovery," lies midway between Cathay and North America. In the Indian Ocean, the incorporation of new data is astonishingly poor, with India and Sri Lanka still in reversed proportions, as per Ptolemy.

The true importance of this map lies beyond its geographic novelties, however. Though geographically dated, it introduces a radical new concept in the understanding of earth's nature and its place in the cosmos. While prevailing cosmological theory placed the earth at the center of the universe with the cosmos orbiting around it, here the earth is shown rotating on an axis, illustrated by cherubs at either pole churning the planet around with cranks. As this clear reference to axis rotation precedes the publication of Copernicus' *De Revolutionibus Coelestium* by eleven years, the map's author no doubt learned about Copernicus' principles either orally, as they were shared by word of mouth for two decades prior to the publication of the book, or from one of the copies of his brief manuscript work *Commentariolus,* which Copernicus is known to have circulated among selected friends by 1529.[194]

The cranks with which Holbein's cherubs allegorically spin the earth are themselves of interest as an early record of this still relatively recent innovation. The development of the device about a century earlier marked a major advance in late medieval technology.[195]

The fantastic figures surrounding the map represent people and fauna from various parts of the world. Some of these, such as the Ubangi people of Africa, are quite real; others are the invented descendants of medieval myth. Taken as a whole, the map is a pastiche of elements spanning the classical (Ptolemaic Indian Ocean), medieval (mythological figures), late medieval (crank, Polean geography), and Renaissance (American and African geography), as well as fresh scientific theory of the underground radical fringe: the as yet unpublished

194. Copernican theory regarding the workings of the universe so challenged Church doctrine that it was first circulated only in manuscript in his book <u>Commentariolus</u>. It was not until Copernicus was near death, in 1543, that <u>De Revolutionibus Coelestium</u> was published. Such was the Church's distaste for Copernicus' principles that in 1616 they were formally denounced, and late in 1632 a book being prepared by Galileo was banned before its publication because it supported Copernicus' ideas. Early in 1633 Galileo was forced by the Inquisition to renounce the "philosophically false" view of a non-geocentric universe. Although the Church, under Pope John Paul II, finally "pardoned" (!) Galileo in 1979, it still maintains Galileo's "guilt" at publishing date for this volume. A Vatican statement actually charges that his ideas "have led us to the catastrophe of Marxism and brought us to the end of industrial civilization" (-International Herald Tribune, September 18, 1991).

195. The bit and brace had not appeared until the 1420s, and the double compound crank and connecting rod not until about 1430; Holbein's map predates any theoretical discussion of the device, the earliest known such work being a book by Giuseppe Ceredi not published until 1576. See Dallas Pratt's article "Angel Motors," in <u>Columbia Library Columns</u>, May, 1972.

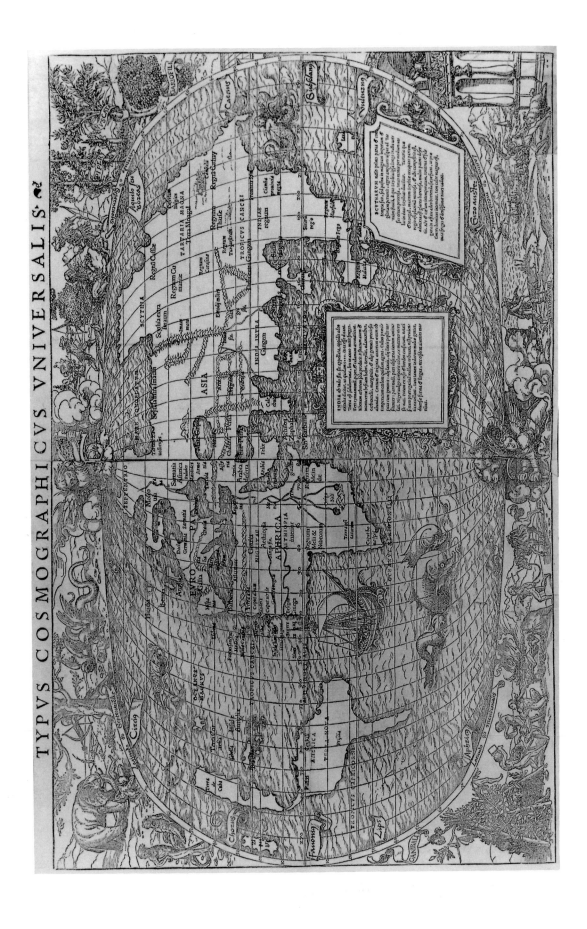

PLATE 14   anonymous / Hans Holbien, Basle, 1532   (entry 21)

theory of axis rotation, an alien and disorienting concept which thrust upon the map's viewer the ultimate culture-shock : that of challenging one's long coveted status in the universe.

- - - - - - - - - - - - - -

22. THE WORLD

[untitled].

Antonio Florianus, Venice, circa 1555 (after the 1538 map of Mercator). From an Italian composite atlas.

Medium : copperplate engraving.
Size of original : 460 x 835 mm.

PLATE : 15

This map of the world by the Italian painter and architect Antonio Florianus[196] employs a novel projection in which the earth is presented in polar hemispheres, each divided into fully 36 gores. Florianus may have used the projection in an attempt to minimize distortion, or to allow it to be dissected and mounted as a globe. Based on his request to the Doge of Venice to obtain a privilege for the map, it seems that Florianus intended it to serve both purposes:

> " [W]*ith my diligence and knowledge having made a mappemonde which has never been made before, with the aid of which one can easily study and learn cosmography and see the entire picture of the world, since it can be reduced to spheric form as Your Highness can ascertain.*" [197]

Contrary to Florianus' boasts, however, neither his geography nor his projection were original. The clever projection was taken from a world map of Alonzo de Santa Cruz (manuscript, 1542).[198] And as for his geography, Florianus copied a double cordiform world map of Antonio Salamanca (Rome, circa 1550), itself an unacknowledged plagiarism of a map of Mercator (Louvain, 1538).[199] Minor spelling variations introduced by Salamanca have been carried over by Florianus, who also introduces a few errors of his own.

One example of this is Antilia, the mythical island of Seven Cities, which lies just below 50° north latitude in the Atlantic. Florianus converts the Arabic numerals used by Mercator and Salamanca into Roman numerals, but in doing so he accidentally robs Antilia of one of its precious cities : *Insulae VI civitatum.*

To his credit, Florianus does attempt to add new data in South America, where he charts a river called *R. Maragnon* which is not shown by Mercator or Salamanca. This is probably the Amazon, borrowed from Gastaldi's map of 1546. He also happily spares his audience the absurd reference to a kingdom of parrots *("Psitacorum regio")* which both Salamanca and

---

196. Records survive showing that Florianus was hired to make such items as an altarpiece, paintings for city nobles, and cantinels (small painted wooden boards which were secured to the ceiling between supports).

197. Rodolfo Gallo, "Antonio Florian and his Mappemonde," Imago Mundi VI, p. 35.

198. Santa Cruz assisted in the padron general under Charles V. See Nordenskiöld Periplus, plate 50.

199. Mercator's map, in turn, was largely based on the 1531 map of Fineaus (entry 19).

PLATE 15   Antonio Florianus, Venice, circa 1555   (entry 22)

Mercator had located in *Terra Australis*. This "parrot land" was, of course, the Brazil of Cabral.[200]

America is correctly separated from Asia, with a large "Oriental Ocean" between the two. The absence of an "Amerasian" landbridge allows Japan to be placed much closer to its true northern latitude than on contemporary maps which connected the two continents. The single landmass which forms the Arctic region is a much more accurate rendering of the true Arctic than the four-island fancy used by Ruysch (1507) and later to be adopted by Mercator himself. Significantly, the name *America* is now used to denote both continents rather than just South America.[201]

Two panels evidently designed to contain text are blank, as are two of four circular frames in the corners. The upper two circular frames contain portraits of Ptolemy and of Florianus.

- - - - - - - - - - - - - -

200. See pages 71-72 in entry 19 (Finaeus).

201. This development is first found on the so-called "Paris Green" or "Quirini" globe of 1513-15.

Chapter VI

TRULY A FOURTH PART OF THE GLOBE

23. AMERICA

Nova Insulae, XVII Nova Tabula.

["New Islands, the 17th New Map"]. [Map of America in *Geographia universalis, vetus et nova, complectens Claudii Ptolomei Alexandrini enarrationes Libros VIII. Quorum primus nova translatione Pirckheimheri. . .*] [colophon: *Basilae apud Henricum Petrum mense Martio, An. MDXLII.*] Sebastian Münster, Basle, 1540 (1542).

Medium : woodcut.
Size of original : 275 x 340 mm (including title).

PLATE : 16

Closely following the New World's debut as a single, unbroken and fully-formed independent continent by Mercator (see entry 22), this coarsely executed woodcut work is the first map of the Western Hemisphere as such.[202] Significantly, Münster has brandished the telling term *NOVUS ORBIS* over the entirety of America, rather than just South America, thus finally freeing North America from its obsolete role as a new *land* on an old *world*.[203] But despite this milestone, the map's title, "New Islands" —— separately set and not an integral part of the woodblock —— still clutches to the stubborn notion that America is an oceanic archipelago, in effect that the "old world" comprised the world's only continents. Another attempt to regiment the New World into the Procrustian bed of ancient beliefs is found in the map proper, where the phrase *Insula Atlatica quam vocant Brasilii & Americam* identifies America as the lost Platonic island/continent of Atlantis.

With the voyage of Magellan (1519-22), the southerly extent and magnitude of the New World was fully realized, and America had proven itself to be a formidable impediment to those hoping to reach the Orient. Although no serious hope remained of finding a shorter route around South America than that found by Magellan, some sought a more practical passage through North America.

In was in the hope of finding such a route that in 1524 Francis I sent the Florentine explorer Giovanni di Verrazano to sail up the eastern seaboard in a vessel named *Dauphine*. Verrazano, it seems, was overly eager to believe he had succeeded in his mission: when he passed along the Outer Banks between Capes Lookout and Henry, he mistook those banks to be an isthmus of North America tenuously separating the Atlantic and Pacific Oceans. Verrazano related that he

---

202. A more elementary hemisphere is found on the right inset of the 1507 world map of Waldseemüller, and the 1512 copy by Stobnicza. In that earlier configuration, however, North America is not fully developed, is nearly but ambiguously contiguous with South America, and the southern extent of America is left to the viewer's imagination.

203. A subsequent (and more common) state of the map deletes "NOVUS ORBIS" covering North and South and substitutes "Die Nüw Welt" and "Novus orbis," both confined to South America alone.

"found an isthmus a mile in width and about 200 long, in which, from the ship, was seen the oriental sea between the west and north. Which is the one, without doubt, which goes about the extremity of India, China and Cathay. We navigated along the said isthmus with the continual hope of finding some strait or true promontory at which the land would end toward the north in order to be able to penetrate to those blessed shores of Cathay." [204]

Verrazano's "oriental sea" was nothing more than the Pamlico and Albemarle Sounds. While history has been sympathetic to this colossal error, it has failed to explain Verrazano's bizarre response to it. After claiming to have unearthed the "holy grail" of navigators, Verrazano apparently made no attempt to cross the "isthmus," either on land or through any of the inlets he reported, inlets through which one of his longboats would have had little difficulty maneuvering. Nor does Verrazano appear to have made an attempt to return to his promising isthmus.[205] What *can* be explained is the failure of Francis I to follow up Verrazano's supposed gateway to the Orient. By the time Verrazano returned, Francis I was a prisoner of the Emperor, Charles V, a long-time rival who apparently defeated Francis' own bid to be Holy Roman Emperor through vote-buying. Although Francis I was released from prison upon signing the Treaty of Madrid (1526), he was obliged to give two of his sons as hostages to insure his compliance with its terms; once free he declared the treaty void, but it would clearly have been foolish for him to send yet another expedition to the Southeast, as those shores were "legitimately" claimed by Spain.

By the end of that decade some cartographers had adopted the bogus isthmus, and it was popularized through the printed book with this map of Münster.[206] A Verrazanian passageway through the mid-Atlantic coast was sought as late as the Jamestown settlement of 1607.

Münster chokes the mid-Atlantic coast of North America to accomodate the isthmus, creating a huge sea cutting through Canada. The upper terrace of this two-tiered continent, which includes all of the eastern seaboard from Virginia and above, is named *Francisca* for France and Francis I. Britain's early expeditions out of Bristol are acknowledged by *C. Britonum*. The island of *Corterati* is the Newfoundland of the Corte-Real's, and Spanish dominance in the southeast is evident from *Terra florida,* the only North American place-name south of the isthmus.

The Iberian superpowers' real sphere of influence lies further south. A Portuguese standard off the coast of Brazil and a Spanish standard planted in Puerto Rico tag the New World according to papal decree and signal the beginning of the colonial epic.

In the Caribbean, above the "island" of Yucatan,[207] are two islands marked *ins. Tortucaru,* the Tortuga Islands. These islands, found by Ponce de León while attempting to return to Puerto Rico from Florida, are so named because de León's men claim to have captured 170 tortoises there in a single evening. Due west of the Tortugas, just off the coast of Mexico, lies *Panuco,* which like the Yucatan is mistakenly shown by Münster as an island. Pánuco was a coastal settlement on the Mexican mainland which had marked the Gulf threshold of Montezuma's empire, and became a pivotal point in rival Spanish claims. In 1521

---

204. Quoted from Cumming et al., The Discovery of North America. The direction "west" was corrected from the original text's "east."

205. Verrazano's original report is not known to survive; copies were circulated in Italian, and one was printed by Ramusio. Verrazano himself is believed to have made another voyage to America and is reported to have been killed and devoured by Brazilian Indians.

206. Along with the world map from Münster's atlas. Verrazano's pattern was followed in the manuscript maps of Vesconte Maggiolo (1527; destroyed in World War II), Girolamo Verrazano (1529), and the globe of Robert de Bailly (1530). Evidence of some Verrazanian influence is visible on the anonymous printed world maps of circa 1530 (Shirley #62) and circa 1535 (Shirley #71).

207. The Yucatan was generally believed to be an island throughout the first half of the century. See footnote 150.

·NOVAE INSVLAE XVI·XVII·NOVA TABVLA·

PLATE 16

Sebastian Münster
Basle, 1540

(entry 23)

it was cited in a royal cédula as the southwestern limit of a territory granted to Francisco de Garay which extended to Ponce de León's Florida, and so became the site of decisive confrontation between Garay's men and his rival, Cortés. To the southeast of the Yucatan is *Cozumela,* an early mapping of Cozumel Island, whose people *"goe naked* [and] *knew not to what use war serves . . . They are poore people, but very charitable and loving in their false religion and beliefe."* [208]

Catigara, a realm from the Africa-Asia landbridge of Ptolemy, lies along the coast of Münster's Peru. Assuming that the northern segment of the landbridge was Indochina, Cattigara occupied the approximate spot of what is now Hanoi. It probably migrated to Munster's Peru via an earlier map depicting South America as an Asian promontory, where use of that nomenclature was a natural transition, such as the maps of Finaeus (entries 19 and 20). In Brazil is a macabre vignette captioned *Canibali.* South of this cannibalism reference lies *Regio Gigantum,* purported kingdom of giant "Patagonian" people reported by Pigafetta from Magellan's voyage. Some literate member of Magellan's crew appears to have been reading the then-popular story *Primaleon of Greece,* published in Castile in 1512, in which a giant, the Grand Patagon, was brought home by the hero as a trophe (Magellan attempted to do the same, but his *patagon* did not survive the return voyage).

To the west of Magellan's strait appears the name *Mare pacificum.* This is the first time that a printed map referred to the great ocean by its modern name, the name which Magellan had christened it.[209] The dominant influence in the Pacific, however, is still Marco Polo. Most prominent is Polo's *Zipangri* (Japan), still two or three years before its earliest known contact with Europeans, and hence mapped arbitrarily as a rectangular, north-south oriented single island.[210] Because an extensive archipelago has been placed between it and the Asian mainland, and because the Pacific itself is so undersized, Japan hovers just off the "California" coast. The island complex between Japan and Asia is Polo's *Archipelagus 7448 insuaru.* Marco Polo related that

> *"according to the testimony of experienced pilots and seamen that sail upon* [the China Sea] *and are well acquainted with the truth it contains 7,448 islands, most of them inhabited."*

These are probably the Philippines, actually comprised of approximately 7100 islands. If so, the fact that Asian pilots by Polo's time could have so accurately inventoried such a vast island complex reveals the extent to which their navigation and trade had by then developed.[211] Polo assures the reader that

> *"in all these islands there is no tree that does not give off a powerful and agreeable fragrance and serve some useful purpose . . .* [and there

---

208. Gómara, <u>The Conquest of the Weast India</u>, p. 35-36.

209. This map, and Münster's world map from the same work, were the first printed maps to use the term "Pacific." The name was used by Pigafetta in a manuscript map of the Strait of Magellan made circa 1525.

210. The earliest known European encounter with Japan occurred in 1542-43. Matsuda Kiichi, the Japanese historian, accepts the date of September 23, 1543, as that which saw three seamen, having approached the Orient from Portugal via Africa, blown off course to a chance landing on Japanese soil. Offical Portuguese secrecy may have hidden earlier encounters.

211. Variant copies of Polo specify 7,459. This author favors their identification as the Philippines, but the Indonesian islands are also a possibility. It also can be argued that these islands are simply part of the medieval island-mythology rather than any real archipelago, but in context it seems that Polo was simply relating what he himself had been told by his Chinese hosts, and it appears in Polo's text with the "true" report about Japan. Chinese, as well as Arab, Javanese, and Japanese merchants and adventurers had all reached the Philippines by Polo's time; the earliest recorded reference to them is from an Arab ship calling at Canton in 982 A.D. (see Quirino, <u>Philippine Cartography</u>, p. 3). Münster's map of Asia in the same atlas shows both the 7,448 islands and one "true" Philippine island, Puloan, one of the archipelago discovered by Magellan in 1521.

are], *in addition, many precious spices of various sorts . . . pepper as white as snow and in great abundance, besides black pepper."*

He goes on to explain that the voyage from Zaiton or Kinsai to these islands, though enormously profitable, takes a full year to complete, for only one wind blows towards them (in the winter), and only one blows back to China (in the summer). And the ocean in which they lie, though called the "China Sea," is really *the* Ocean, in other words the encircling Ocean Sea, preserved most clearly on the T-O map of Isidorus (entry 1).

The ship illustrated in the Pacific is the *Vitoria* of Magellan. Directly below it are the two *infortunatae* islands whose foreboding name typified that expedition's austere maiden Pacific crossing. With much of its crew already suffering from hunger and scurvy, the expedition

"discovered two little Islands uninhabited, where they sawe nothing but birds and trees, and therefore named them infortunate Islands." [212]

Magellan, like those who followed in his path, missed entirely the major fertile Pacific island groups which would have provided periodic oases of sustenance.

Straddling the western end of the map is a land named *Calensuan,* one of the islands which Magellan's surviving crew reached after his death in the Philippines. The island was reported to be inhabited by "Moors" who had been banished *"owt of the Ilande of Burnei* [Brunei /Borneo]*."* [213]

Sebastian Münster was raised as a Franciscan monk, converted to Lutheranism, taught Hebrew at Heidelberg and Basle, and was proficient in Greek and some Asian tongues. He died of the plague in 1552. First published in 1540, his atlas was the first to contain separate maps of each of the four continents.

- - - - - - - - - - - - - - -

212. Marytr/Eden Decades, p. 221.

213. Marytr/Eden Decades, p. 228. Magellan's Borneo can be found on Münster's map of Asia as "Porne."

24. THE WORLD

Carta Marina Nova Tabula.

["A new sea chart {of the world}"]. Giacomo Gastaldi, Venice, 1548.
From Gastaldi's edition of Ptolemy's *Geographia.*

Medium : copperplate engraving.
Size of original : 135 x 175 mm.

PLATE : 17

This sea chart of the world depicts the entirety of the continental northern landmasses as an unbroken ring around the globe. North America and Asia form a single mammoth continent, which in turn connects to northern Europe via Greenland. The Asia-America connection was a standard concept, and one of which Gastaldi was a particularly strong endorser. This map's linking of North America and Europe is highly unusual, however, but was a natural consequence of two errors: on the east, Gastaldi depicts Greenland as an elongated east-west outgrowth of Scandinavia, a perculiar pattern used by Waldseemüller earlier in the century (entry 11); on the west, he adopts the Verrazanian model for North America which had been sanctioned by Münster in 1540 (entry 23). In combination, these two flawed elements stretched out over the North Atlantic and, quite logically, joined.

The implications of Verrazano's geography were exciting. An enterprising merchant might simply construct a vessel on the far (western) shore of the isthmus (which Verrazano believed to be as narrow as a mile in width) to conduct an easy two-stage rendevous with China, thus finally succeeding in establishing the viable trading route to the East which Columbus and so many others had sought. At the far end of this journey, crossing the "China Sea," Gastaldi shows *Tangut,* the place in China where the messengers of the Kublai Khan had met Marco Polo. At the bottom of the isthmus, the junction between perceived Asia and perceived America, Gastaldi has marked *montagna verde,* an early reference to the Appalachian (Blue Ridge?) Mountains.

On the Atlantic coast by this "green mountain" is a large, unnamed cape pointing upwards. This cape, which appears to have made its printed debut on the "Ramusio" map of 1534 and appeared in manuscript as early as the Ribero chart of 1529, is found on such later works as the de Jode North America (entry 41). It is often presumed to be Cape Cod, and indeed both the Ribero and Ramusio prototypes reflect the reconnaissance along the New England coast of Estavão Gomes, who is believed to have scouted that cape. But on both those maps, the closest original sources we have, the cape appears on the more southern landfalls of Ayllón rather than those of Gomes' New England. De Jode later mis-matches it to *C. de las Arenas,* which, though indeed a Gomes place-name, more likely represented Cape May.

While the influence of Marco Polo is evident in the parts of Asia which join North America, the old Polean bonds have finally been severed in Southeast Asia and the Australasian islands. A "modern," if still inaccurate, Sumatra and Java have replaced Polo's "Java Minor" and "Java Major," and the "true" Singapore and Malacca Straits now appear, unnamed on the present map because of space constraints but with an early reference to Singapore *(Cinca Pura)* found on the regional map from the same atlas. On the north of the Malay Peninsula, Burma (or, for the moment, *Myanmar*) appears by its modern name *(berma).* Above *berma* is an extremely early appearance in print of the term "China" *(LA CHINA),* with the old Cathay *(CATAYO R)* retained but relegated to the approximate region of Tibet.

In Europe, one curious feature is worth noting. The northwest tip of Spain is pushed out in a stylized fashion, clearly marking it as the most westerly point in Europe (which it is not). This is *c. finis terre,* the "cape of the end of the earth," a little peninsula which had been considered the westerly end of the world by the Romans. That Gastaldi chose it as his only named physical feature in Western Europe, and that he distorted the Iberian coastline to stress its meaning, suggest that, in the midst of his quest for the latest cosmological truths, he allowed himself a moment of sentimentality about a past time that was a different world.

- - - - - - - - - - - - - -

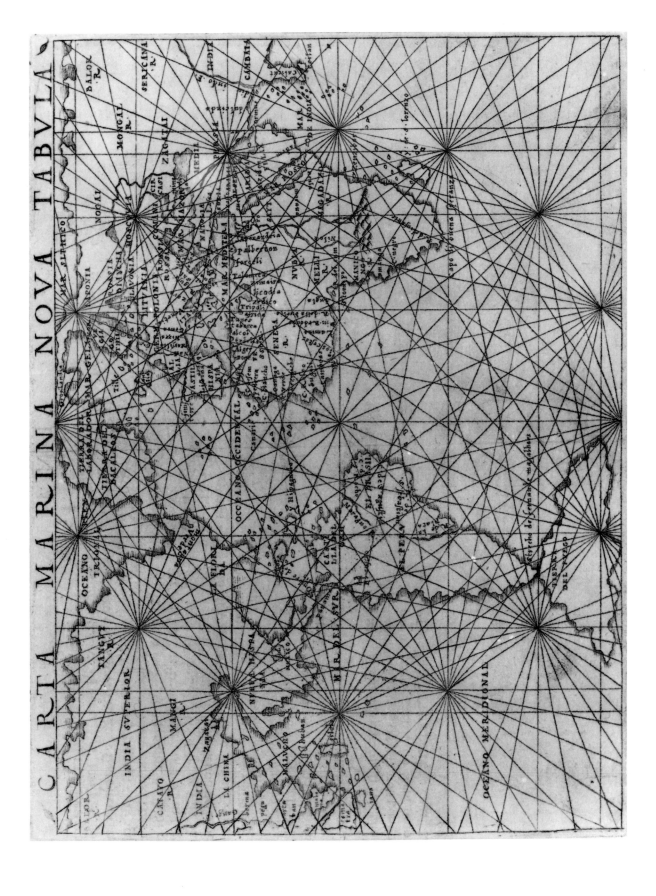

PLATE 17

Giacomo Gastaldi
Venice, 1548

(entry 24)

25. EASTERN SEABOARD [North America]

Tierra Nueva.

["New Land"]. Giacomo Gastaldi, Venice, 1548.
From Gastaldi's edition of Ptolemy's *Geographia*.

Medium : copperplate engraving.
Size of original : 130 x 170 mm.

PLATE : 18

The voyage of Verrazano (1524), made for Francis I of France in search of a passage to the Orient, had provided a preliminary, partial survey of the eastern seaboard of North America. Although the voyage was the origin of much confusion because of its report of a continental isthmus in the Southeast, it did contribute greatly to the mapping of the coast as far north as Narragansett Bay. A decade later, France once again sent an expedition to find a route around or through North America, this time under the command of Jacques Cartier and bound for the region north of that explored by Verrazano. Cartier, whose previous experience in the New World was limited to commerce with Brazil, explored the Gulf of St. Lawrence on his first voyage to North America (1534), and penetrated the St. Lawrence River as far as Montreal on a second mission (1535).[214]

Gastaldi has combined data from the Verrazano and Cartier voyages to create this small work, the first printed map devoted to the East Coast of North America. Fusing the two sources created confusion, however. When Verrazano scouted the East Coast of North America, he cautiously headed out to sea after Narragansett Bay to avoid treacherous shoals, but in doing so missed Cape Cod and much of the New England coast. Cartier, on the other hand, did not venture further south than Cape Breton. As a result, Gastaldi grafts eastern Canada onto southern New England, with Cape Breton Island sprouting directly from Narragansett Bay. Mercifully, Gastaldi ignores Verrazano's famous Outer Banks "isthmus" report and spares the coast the distortion that arose therefrom; unfortunately, though, he does not incorporate data from Cartier's exploration of the St. Lawrence on his second voyage.

Fresh influence from Verrazano is seen in the Northeast, where a new domain makes its debut on a printed map: along with the familiar regions of *Laborador* and *Bacalaos* lies *Tierra de Nurumberg*.[215] Although Gastaldi's spelling suggests that he associated it with the German city, it was more commonly spelled "Norumbega" and was probably derived from an Algonquian Indian word which meant something on the order of "quiet place between two rapids."[216] The term soon came to denote the region of New England. Further south is New York, calibrated about 5° too far north, and called *Angoulesme* after Verrazano's designation, *Angoleme*. The flat shore just up the coast from it, near the name *Flora*, is Long Island but without Long Island Sound. Near *Larcadia* are two inlets, probably the Chesapeake and Delaware Bays.

- - - - - - - - - - - - - -

214. A third voyage is known from Hakluyt's Principal Navigations but no log survives.

215. The term originated with the voyage of Verrazano, or at least it first appeared on the portolan made by his brother Girolamo. It appears as "Oranbega" on the Girolamo Verrazano portolan (manuscript, 1529).

216. See Morison, The Northern Voyages, p. 464.

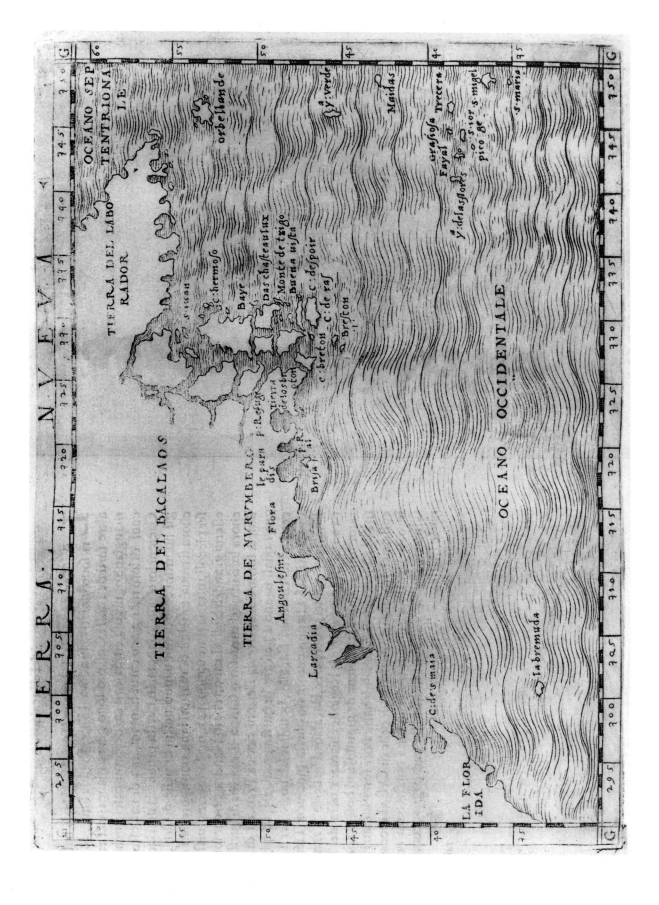

PLATE 18

Giacomo Gastaldi
Venice, 1548

(entry 25)

26. NEW ENGLAND

La Nuova Francia.

["The New France"]. Giacomo Gastaldi /Giovanni Batista Ramusio, Venice, 1556. Map of New England and eastern Canada from Ramusio's *I Navigazioni et Viaggi.* First state of three.

Medium : woodcut.
Size of original : 270 x 375 mm.

PLATE : 19

The chronicler Ramusio credits Giacomo Gastaldi with authorship of this map, and it is in any case clearly a modified extraction of the New England portion of Gastaldi's *Tierra Nueva* of 1548 (entry 25). It is the first printed map focusing on the region of New England, and it established the name "New France" *(La Nuova Francia)* to denote the French possessions in the north.

This work adds some interior geography to the 1548 *Tierra Nueva,* and depicts the coast in more elaborate, if no more accurate, detail. The courses of the several rivers have been modified, and a westerly branch of the Hudson has been added. The landmass called *Isola de Demoni* on the 1548 work is now part of Labrador (extending off the map), but the name has been retained for the northernmost and largest island of Newfoundland. It has been suggested that the demons illustrating this *Isola de Demoni* are the Beothuk Indians, described by Pasqualigo half a century earlier as "shy" and "gentle," now defending themselves against European aggression.[217] Separating *Isola de Demoni* from *Terra Nuova* (i.e., Newfoundland) is the *Golfo di Castelli,* which was the *Bay des Chasteaulx* of Cartier, present-day Strait of Belle Isle. Ramusio resurrects the old *Baccalaos* (cod fish) for one of the other islands to the south. *Bonne viste* is Cape Bonavista, Cartier's first sighting of land in Newfoundland. The illustration of birds nearby reflects Cartier's *Isle des Ouaisseaulx* ("Isle of Birds," now Funk Island). *C desperaza* (i.e., *"de speranza"*) is North Pt. Miscov, and *C. de'ras* is Cape Rouge. A cross with a fleur-de-lis is planted on that island, symbolizing Cartier's claiming of the region for France. The two major islands which separated the mainland and *Terra Nova /Isola de Demoni* on the 1548 work have here been eliminated. In the Atlantic is a stylized representation of the Grand Banks, the mid-point of which is *Isola della rena* ("Island of Sand"). As with the 1548 Gastaldi work, Cartier's penetration of the St. Lawrence to Montreal is not shown.[218]

Angoulesme, the strawberry-shaped bay in the lower part of the map, is Verrazano's New York Bay, which the navigator described as

> *"a very attractive site between two small prominent hills, in between which a very great river flowed to the sea* [the Hudson], *deep at its mouth, and from sea to the place where it merged any loaded ship could go on a rising tide, which we found to be eight feet. Having anchored off shore in a sheltered place we did not wish to venture farther without knowing the nature of the river mouth. We took the longboat from river to the land, which we found greatly inhabited. The people were about the same as we had met before, dressed with bird feathers of different colors, and came toward us happily, giving loud shouts of admiration, and showed us where we could take the boat safely."*[219]

---

217. See Ganong, <u>Crucial Maps . . .</u>, p. 271, and Coolie-Verner, <u>The Northpart of America</u>, p. 12; see pages 56-57 of entry 14 (Sylvanus) for Beothuk Indians.

218. Although Ramusio included that voyage, along with a plan of Montreal, in the third volume of the <u>I Navigazione</u>, the same volume in which this map appeared.

219. The Verrazano quotes are from Sauer, <u>Sixteenth Century North America</u>.

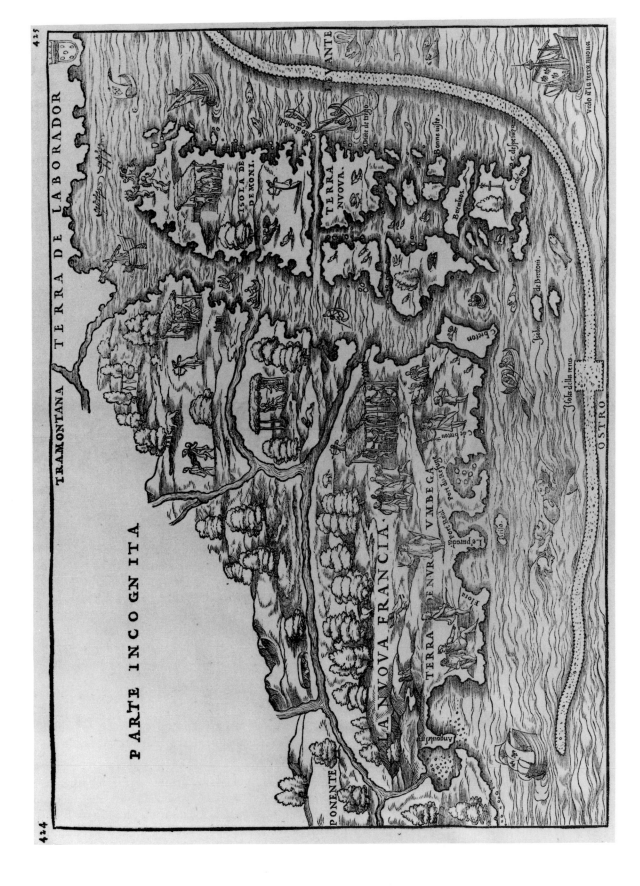

PLATE 19

Giacomo Gastaldi/
G. B. Ramusio
Venice, 1556

(entry 26)

The longboat ventured as far as Upper Bay, but then suddenly *"there arose a contrary wind from the sea which forced us to return to the ship and, greatly to our regret, to leave that land."* Verrazano thus missed the opportunity to enter the Hudson (though Ramusio speculatively charts it).

East of New York is the flat, east-west oriented southern coast of Long Island (lacking Long Island Sound), which Verrazano skirted *"always in sight of land."* The fin-shaped eastern shore of Long Island is discernable at *Flora*. Verrazano continued east to Newport Bay *(Port Real)*, *"a most beautiful harbor* [where upon entering] *we saw about twenty boats of people who gathered about the ship with various cries of astonishment."*

Finally, *Port du Refuge* is Narragansett Bay, in which *"a fleet of any size could stay . . . in security without fear of tempest or other hazard of fortune."* After this point, Verrazano's reconnaissance becomes cursory, and his influence on Ramusio ends.

Clipping the coasts of much of what is now Massachusetts and Maine, Cartier's Cape Breton and Cape Breton Island are grafted onto the eastern bounds of Narragansett Bay. Due both to the omitted stretch of coast and to the effects of magnetic declination of the mariner's compass at high latitudes, southern Newfoundland is shown slightly south of New York Bay.

- - - - - - - - - - - - - - -

27. AMERICA

Brevis, exactaq. totius novi orbis evsq insularum descriptio recens A. Ioan. Bellero edita.

["Brief and exact description of the entire New World and its islands, newly rendered by Joan Bellero"]. Map of America by Juan Bellero, 1554, in the work *La Sphere des Deux Mondes, composee en Francois, par Darinel pasteur des Amadis,* Darinel de Tirel, Antwerp, 1555 (the map itself had first appeared in Gómara's *Mexico* the previous year).

Medium : woodcut.
Size of original : 170 x 130 mm.

PLATE : 20

The history of this map, like the previous works, is largely rooted in voyages which considered North America more of an obstacle than a destination. This map, however, draws on Iberian expeditions for its North American nomenclature.

When in 1519 Magellan set sail on his quest for a passage around America, by ill-circumstance the Portuguese pilot Estavão Gomes, a bitter rival of Magellan, was the sailing-master of one of the fleet's ships. While the fleet was diffused exploring their new-found strait (1520), Gomes instigated a near-mutiny, abandoned the expedition, and eventually returned to Spain. The Crown, which had hoped a route would be found into the South Sea at about the latitude of the Rio de la Plata, learned from him that the expedition had already ventured in excess of fifteen hundred kilometers further south. This was far too steep a latitude to invite routine access to the Orient even if the strait yielded the anticipated passage. As a result, a better solution was sought even before the return of Magellan's survivors in 1522. In 1524 Gomes attended a congress of both Spanish and Portuguese navigators held in Badajoz to discuss the likelihood of a passage to the Moluccas existing north of Florida, and the following year he sailed from the northwestern port of La Coruña with instructions from Charles V to search for such a route along the coast of *Bacallaos*. He thus became the first Spanish explorer in New England. The surviving evidence of his voyage shows that he recorded the Maine and Massachusetts coasts including Massachusetts Bay and Cape Cod, filling the gap between the data of Verrazano and Cartier used by Gastaldi in 1548.[220]

On Bellero's map it is possible speculatively to identify Gomes' nomenclature. The *B. d. S. Christoval* at the pronounced indentation in the Atlantic coast is probably Cape Cod and Massachusetts Bay, named by Gomes for Saint Christopher because he reached it on that saint's feast day, July 25. Continuing east, the Merrimac River is *R. d. buena madre* ("River of the the Good Mother"), named for Saint Anne, as Gomes reached there on her saint's day, July 26. *Rio seco* ("Dry River") is the Saco River, *Arcipielago* (i.e., *Archipelago*) is Casco Bay, and *C. de S. maria* is probably Cape Small. *R. de las gamas* ("River of the Deer") would be Penobscot Bay and River. Just east of it is *Costa de medanos* ("Coast of Sand Dunes"), possibly a reference to the sand-like appearance of the mountaintops of Mount Desert Island. *R. de montanas* (River of Mountains) is Frenchman's Bay or Pleasant River, and *Castenal* ("Chestnut Grove") may be Machias Bay. Continuing northeast after a small void area is *B. de la ensenado* ("Bay of the Rounded Gulf"?), which may correspond to Passamaquoddy Bay. *R. de la buella* (i.e., *vuelta*, or "River of the Return") is probably the River St. John, from where Gomes began the voyage back to La Coruna.[221]

Another man whose labors provided some of Bellero's information, and whose ambitions at least partly targeted the finding of a passage through North America, was Lucas Vasquez de Ayllón, an auditor of St. Domingo. In 1520, while Gomes was making his way back to Spain after abandoning Magellan's fleet, Ayllón obtained a license to explore the coast of North America. He charged Francisco Gordillo to undertake the expedition in his name. While en route through the Bahamas, the caravel encountered and joined forces with another Spanish fleet, that of Pedro de Quexos. Reaching Florida, seventy natives were taken and enslaved. As Ayllon had specifically instructed Gordillo to seek friendly relations with the Indians, the act was condemned upon the caravel's return and Ayllon arranged for the Indians' safe return to their homeland. Ayllón, determined to explore and settle, persevered through both political and practical problems that arose. In 1526 he set sail from Puerto Rico with three vessels bearing the seeds of a colony : fully six hundred men and women, including clergymen and doctors, African slaves (here first introduced to North America), and one hundred horses. While sailing along the North American coast, the fleet lost its brigantine, quickly replacing it with a small boat, a *gavarra*. This was the first European-style vessel built in what is now the United States. They eventually settled at Gualdape (in present-day South Carolina), founding a colony they called San Miguel.[222] Bellero places nomenclature from Ayllon's excursions directly below that of Gomes. Beginning below Gomes' *B. d. S. Christoval* is Ayllón's *C. d. Trafalgar, B. del principe,* and *C. d S. Roman* (below a "river of canoes").[223] Neither Bellero nor any other contemporary mapmaker, however, records the colony of San Miguel, probably because it had quickly failed. Ayllon himself perished there from sickness.

The final major source of Bellero's East Coast cartography is Portuguese. By letters patent dated March 13 of 1521, Portugal's King Manoel acknowledged the discovery by one João Alvarez Fagundes of a mainland and islands on the northeast coast of the New World. The discovery, made no later than 1520, was believed to lie south of the Corte-Real landfalls and north of Spanish claims. Surviving evidence is scant and inconclusive, but Fagundes may have scouted Newfoundland and the St. Lawrence Gulf, and that is the way it figured in

---

220. As recorded on the Gastaldi maps of 1548 and 1556 (entries 25 and 26).

221. As analysed by W. F. Ganong in <u>Crucial Maps in the Early Cartography and Place-Nomenclature of the Atlantic Coast of Canada</u> (1964), p. 174-189. Bellero's map falls into Ganong's "Chaves-Santa Cruz" archtype. According to another view of the nomenclature, "B. d. S. Christoval" is lower New York Bay, "R. d. buena madre" is perhaps the Connecticut or Thames River, and "R. de las gamas" is the Hudson River, though misplaced (see Sauer, <u>Sixteenth Century North America</u>, p. 67-68).

222. According to Winsor, however, the Gualdape colony of San Miguel was the same site as Jamestown, Virginia (see <u>Narrative and Critical History of America</u>, vol. III, p. 241).

223. The remaining place-names in the Southeast on Bellero's map correspond to those on the Spanish "padron general;" see e.g. Harrissee, <u>The Discovery of North America</u>, p. 635.

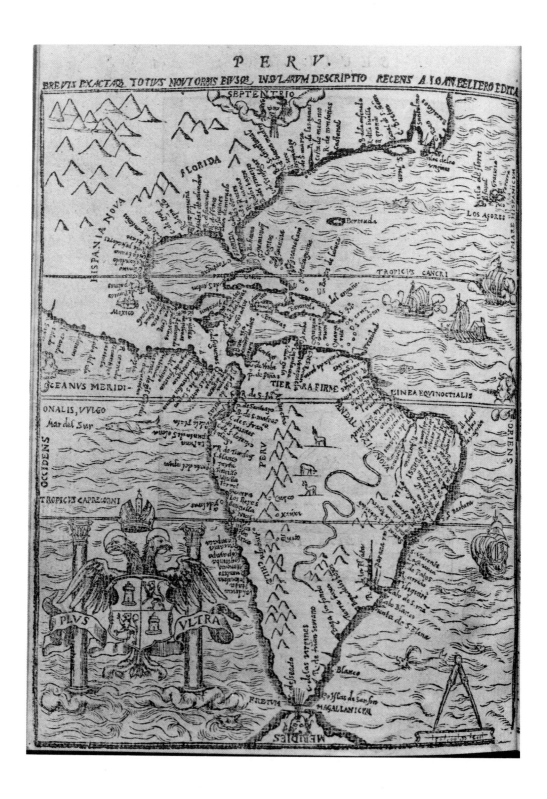

PLATE 20   Juan Bellero, Antwerp, 1554   (entry 27)

Bellero's map. If so, *Santolino* would be the St. Lawrence more than a decade before Cartier. Nearby, *C. Rasso* is Cape Race, and Fagundes' *islas de las virginas* is named for the story of St. Ursula.[224] Following Fagundes' initial exploratory voyage (which he himself may or may not have accompanied), he is believed to have founded a settlement, the first post-Columbian European colony on the North American continent, planned after the fashion of the Azores. According to proposals it was to be a peaceful settlement and engage in the production of soap.[225]

Bellero's depiction of South America resembles that found in the 1546 world map of Gastaldi, both in its accuracy and in the peculiar orientation given the Amazon River, originating in mountains far to the south and flowing north.[226]

This map was published in Antwerp, where Spanish censorship was at this time relatively lax. The book *La Sphere des deux mondes* is a geographic treatise in poetry; it accompanies each of its several woodcut maps by an eight-line cosmological stanza. The one accompanying the map of America waxes lavishly of the country's grandeur and compares its riches to Midas :

O Ultre la mer, se voit le grand Peru
Large estendu, ce semble un autre monde,
D'ou l'or massif est si trestant venu
Que lon fauldroit en somme si profonde.
Mais qu'en est il? le territoire abonde
Mille Midas y vont mourant de fain,
Tel a trop plus, mais encores il sonde,
Et va querant ce quil tient en sa main.

> ("Oh across the sea lies the great Peru
> Greatly extended, it seems like another world,
> where a mountain of gold . . .")

- - - - - - - - - - - - - -

---

224. See page 45 in entry 12 (Ruysch) for legend of St. Ursula.

225. The location of the Fagundes colony is uncertain. Sauer (who believes that the original Fagundes voyage explored the coast of Maine through Sable Island) makes a strong case for the Annapolis lowland on the Bay of Fundy as the site of the colony. Basque sailors trading with the Azores later reported the colony to be faring well and to have established friendly relations with the Indians. What ultimately became of it is unknown. See Sauer, Sixteenth Century North America, p. 47-51.

226. A south-to-north orientation of the Amazon was shown earlier, e.g., Monachus (circa 1527) and Vopell (1536), but Bellero's is most similar to that of Gastaldi.

28. THE WORLD

> [untitled in copperplate; a partially obscured title in manuscript reads] :
> *Universalis De Terrao Orarium ex vera recen* [. . .] *traditione.*

> ["A Map of the Whole Earth according to accurate recent knowledge"].
> [anonymous] (Giorgio Callapoda?), probably Italian (Florence or Venice?),
> circa 1550. Separately published map of the world.

> Medium : woodcut.
> Size of original : 205 x 290 mm.

> PLATE : 21

The author of this map is unknown. Tradition has given it the name *Florentine goldsmith's map* after a listing in a nineteenth century bookseller's catalogue.[227] It closely resembles a map contained in a 1552 manuscript atlas of Giorgio Callapoda, a Greek chart-maker who lived in Crete. However, the fact that the map is not similar to Callapoda's other works, and (though less compelling) the fact that the right half of Callapoda's manuscript is devoid of nomenclature while the printed map has many place-names there, both suggest that Callapoda copied the printed work, rather than vice-versa.

The map shows the world on an oval projection, which had been introduced by Francesco Rosselli about 1508 but had not won a wide audience until used by Bordone (1528), Münster (1532 and 1540)[228] and Gastaldi (1546). This was the first projection which succeeded in depicting the entirety of the earth's surface.[229] The 1546 Gastaldi work initiated a series of related Italian maps using the projection, of which the "Florentine goldsmith" is perhaps the most unusual. As originally engraved, the map largely followed the 1546 Gastaldi work, except that it honored Verrazano's claim of a continental isthmus choking the southeast of North America. This particular example of the map, however, has had Verrazano's isthmus corrected. The western shores of the isthmus appear to have been rubbed from the copperplate itself (rather than from the printed impression), and a new western coastline was drawn in manuscript on the map.[230] Like the Gastaldi map, America forms an unbroken continuation of the Orient.

Balboa's success in penetrating through meso-America to the Pacific in 1513, and Cortés' conquest of Mexico the following decade, both set the stage for Spanish exploration along the heretofore mysterious western shores of America. As a result of these incursions, we now see the California peninsula taking form. In about 1533 an expedition sailing north from *Nueva Hispania* suffered the murder of its captain, Diego Bezerra de Mendoza, by its chief pilot, Ortuña Ximénes, and then, reaching what they believed to be an island, the massacre of Ximenes and many of the crew at the hands of coastal Indians. The survivors of the expedition brought word back to Mexico that the new shores they had chanced upon bore rich pearl fisheries. Possibly enticed by these rumors, in 1535 Hernan Cortés sailed north, reaching a land which he dubbed *Santa Cruz.* Four years later, Cortés sent Francisco de Ulloa north to locate the fabled "Seven Cities," formerly sought on the mythical island of Antilia but by this time thought to lie somewhere in the American Southwest. Ulloa followed

---

227. The bookseller Ellis & White in 1884.

228. See entry 21 for 1532 map; the 1540 Münster which used the oval projection is the world map from his edition of Ptolemy's "Geographia" (as entry 23).

229. Excluding the 1507 gores of Waldseemüller.

230. And so the map's editor appears to have been in possession of the copperplate, possibly preparing to revise it according to this example. The anonymous editor was also the author of the Strabo world map described under entry 4; both works were part of a manuscript cosmological treatise.

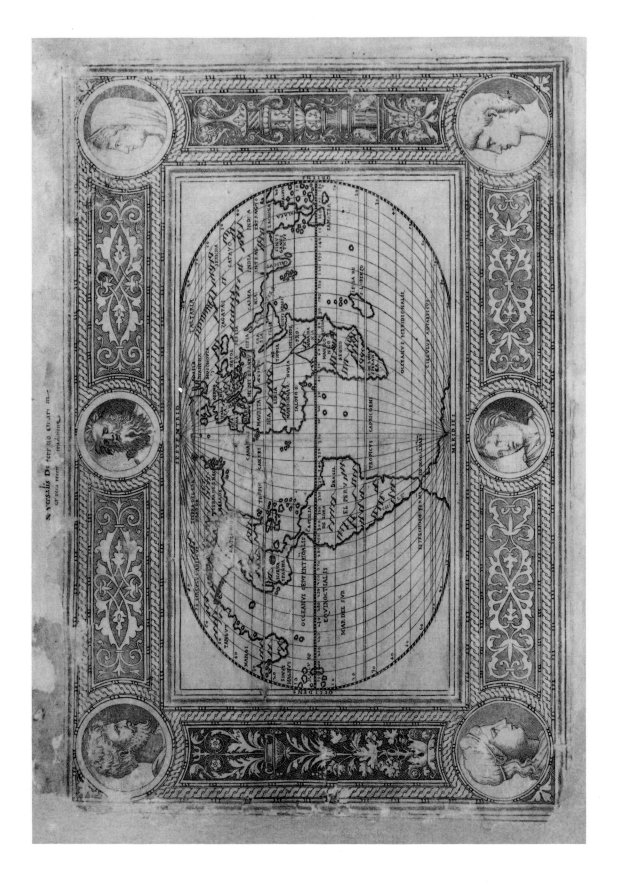

PLATE 21

anonymous/
Giorgio Callapoda (?)
Italian (?), circa 1550

(entry 28)

Cortés' tracks and correctly indicated, if inconclusively, that *Santa Cruz* was a continental peninsula.[231]

Six portraits, of unidentified people, flank the map. Separately published and thereby not afforded the protection of maps bound into books, only three or four examples of this elegant work are known to survive.

- - - - - - - - - - - - - -

29. WESTERN HEMISPHERE

Universale della Parte del Mondo Nuovamente Ritrovata.

["A Map of the Part of the World Recently Discovered"]. Giacomo Gastaldi /Giovanni Batista Ramusio, Venice, 1556. Map of America from Ramusio's *I Navigazioni et Viaggi.* First State of three.

Medium : woodcut.
Size of original : 265 mm diameter.

PLATE : 22

Ramusio states in *I Navigazioni* that this map of the Western Hemisphere is the work of Giacomo Gastaldi. It was the finest general delineation of the hemisphere available to the public at mid-century. Assuming Ramusio's attribution of the map to Gastaldi to be correct, the map also represents an important link in the influential Gastaldi's thinking about the continuity of the American and Asian coasts. Gastaldi, long a steadfast champion of the theory that Asia and America were connected along the northern Pacific, here shows only a tentative, possible link between the Northwest Coast and Asia. Five years after this map, Gastaldi categorically abandoned his long-standing belief in the continuity of the two continents, instead claiming that a strait separated the two.[232]

The northern and southern American continents are labelled *La Nova Spagna* and *El Peru* respectively. Consistent with Gastaldi's earlier maps, the term *America* does not appear at all. The eastern seaboard of North America, though lacking detail, is highly unusual. It exhibits an atypical consistent diagonal orientation rather than the common east-west distortion of the mid-Atlantic coast resulting from magnetic declination.

This is the first printed map to show the Sierra Nevada, the "snowy mountains" noted by Cabrillo in 1542. It is also the first to show the exploits of Francisco Vasquez de Coronado, who led a team of gluttonous Spanish explorers in search of fabulous *pueblos* rumored to lie to the north of their home base of *La Nova Spagna* (Mexico). Closest in their path was *Cibola* (above "La" of *La Nova Spagna*), which Coronado reached in 1540 and which Fray Marcos, a scout sent three years earlier, had described as being grander than Mexico. Cíbola was named either from an Opata Indian name, or from the Spanish *cibola* (bison); like *Antilia* of the previous century, it was believed to consist of seven cities. The pueblo above it is *Cucho,* the Spanish word for fertilizer made from manure and compost. Directly west from *Cucho* is *Tiguas* or Tiguex, the banks of the Rio Grande and home of the Tigua Indians.

---

231. The earliest known printed map to show the California peninsula is Sebastian Cabot's world map of 1544 (one extant example). The "Florentine goldsmith" map is assigned a date of "1544?" by Samuel Hough in <u>The Italians and the Creation of America</u>, entry 106, but it is more likely post-1546 because of its probable ancestry in the Gastaldi work of that year, and pre-1552 if it was indeed the source for Callapoda's manuscript map.

232. See page 101 in entry 30 (Forlani /Zaltieri).

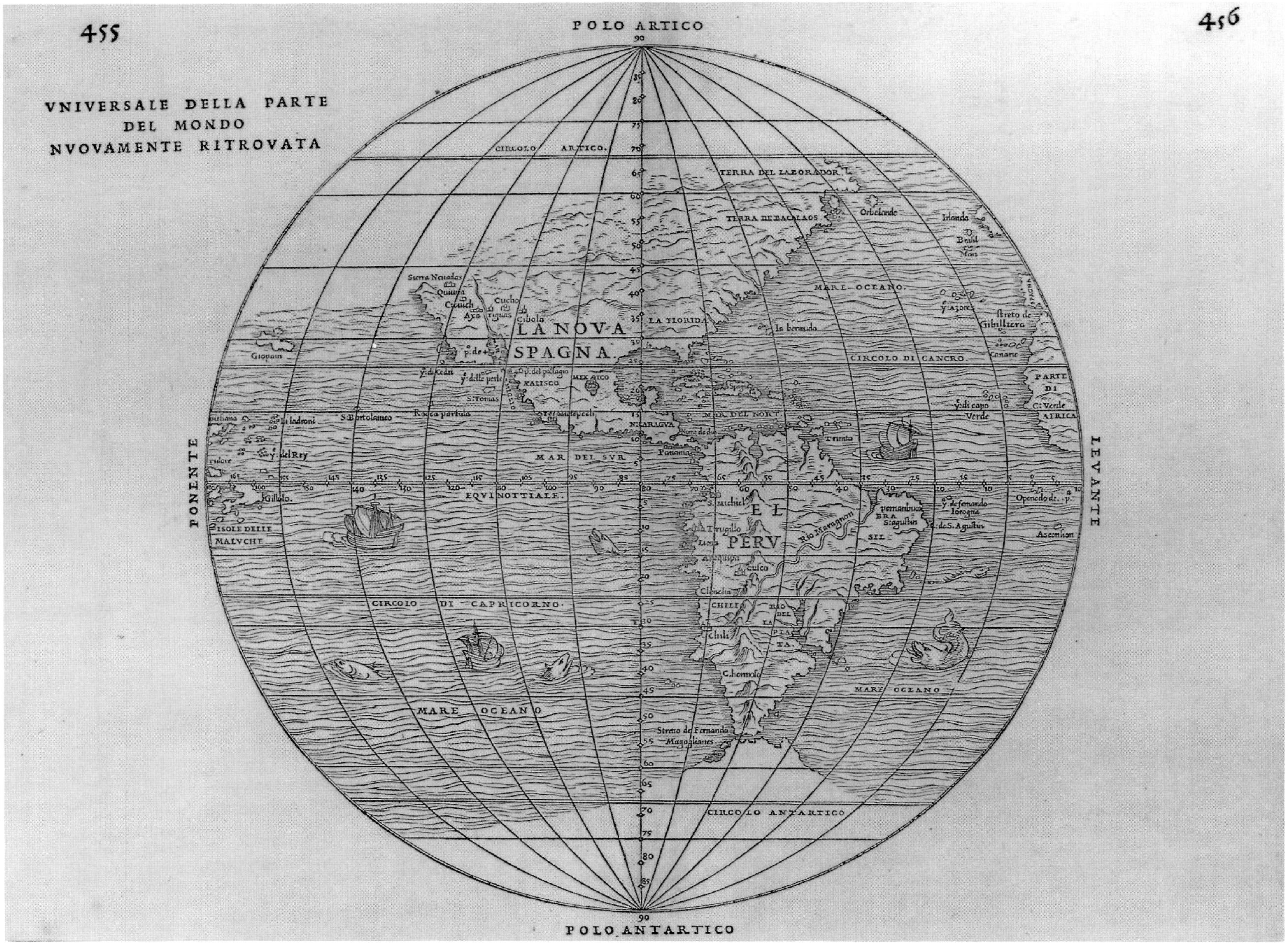

455

VNIVERSALE DELLA PARTE
DEL MONDO
NVOVAMENTE RITROVATA

POLO ARTICO

POLO ANTARTICO

LEVANTE

PONENTE

CIRCOLO ARTICO.

CIRCOLO DI CANCRO

EQVINOTTIALE

CIRCOLO DI CAPRICORNO.

CIRCOLO ANTARTICO

TERRA DEL LABORADOR

TERRA DE BACALAOS

LA FLORIDA

LA NOVA SPAGNA

EL PERV

NICARAGVA

MAR DEL SVR

MARE OCEAN O

MAR DEL OCEANO

MAR. OCEANO

PARTE DI C. VERDE

AFRICA

Irlanda

Ociolande

strito de Gibilterra

C. Verde

ISOLE DELLE MALVCHE

Giappan

PLATE 22

Giacomo Gastaldi/
G. B. Ramusio
Venice, 1556

(entry 29)

This land of a dozen villages was first reached by Hernando de Alvarado and subsequently used by Coronado as winter quarters on his way to Quivera. To the west lies *Axa* or *Haxa*, a village which Coronado had heard about from Indians he called *Querechos*, but which a reconnaissance party failed to locate. Above *Axa* is *Cicuich* (or *Cicuic* or *Cicue*), Coronado's name for the *pueblo* in New Mexico later known as *Pecos*. Between it and *Sierra Nevadas* is *Quivera*, home of the Wichita Indians, the end of both Coronado's journey and of his dreams for riches. Gastaldi shows Coronado's itinerary progressively erring to the west, so that *Quivera*, in what is now Kansas, lies well within California. This was a result of an impression shared by Gastaldi and his contemporaries that Coronado penetrated the continent to the northwest, rather than the northeast. The misunderstanding was reinforced by vagueness in the text of Ramusio, which this map accompanied.

On the southeast coast of the California peninsula, *p. de* ✚ preserves Cortés' original name for Baja California, *Santa Cruz*. This is the site of present-day La Paz. Immediately to the south, just below the Tropic of Cancer, is a small "island of pearls" *(y. de le perle)*. This is a vestige of the earliest insular view of California, of the fact that the tip of California was originally believed to be an island which was lavishly endowed with pearl fisheries. In the seventeenth century California would return to an island configuration, but claiming most of the western coast rather than just a diminuative island.[233]

The gulf separating California from the mainland is designated *Mar Vermiglio* after the Spanish *Mar Bermejo* ("red sea"), so named by Ulloa in 1539 for the reddish color produced by sediment from the Colorado River. In his account of the expedition, which had been ordered by Cortés to investigate the land of *Santa Cruz*, Ulloa commented that they named the gulf *"Ancón de San Andres and Mar Bermejo, because it is of that color and we arrived there on Saint Andrew's Day."* [234]

This appears to be the first printed map to name Japan by its modern name *(Giapam)*.[235] The use of the term does not, however, mean that Gastaldi's primitive Japan is based on post-"discovery" sources, as the term *Japan* reached Europe independently and can be traced back as early as 1513. Nonetheless, Japan does now bear a far more correct orientation than that found on earlier maps. The Portuguese sailors who brought Europe some form of the term *Japan* from their Malay colleagues probably also learned something of the country's basic shape and orientation. Additionally, direct intercourse between Portuguese and Japanese merchants, via trading posts established on islands off the China coast, had begun as early as 1533.[236]

- - - - - - - - - - - - - -

233. A re-cut woodblock which produced states II and III of this same map records the island of pearls as a second, "ghost" southern shore of the peninsula. This was probably simply the result of sloppy copying on the part of the woodcutter.

234. See Polk, The Island of California, p. 143. Some early chroniclers (e.g., Wytfliet) and modern historians (e.g., Polk) believe that the name was meant to suggest a physical similarity or correlation between the California Gulf and the (Arabian) Red Sea.

235. Donald Lach, Japan in the Eyes of Europe (p. 652-653), quotes E. W. Dahlgren, who wrote in 1911 that "Giapam" first appeared on a Gastaldi map of 1550; this appears to have been a mis-dating for Gastaldi's 1561 map of Asia.

236. The term "Jampon" appeared in the Suma Oriental of Tome Pires, which can be dated as early as 1513. Barros refers to "Japoes" in his second "decade" of 1549. Portuguese pilots are believed to have got the name "Japan" from Malay traders. The term is believed to be a Portuguese adaptation of the Malay word "Japun" or "Japang," which in turn was derived from the Chinese "Jih-pên kuo" ("land of the rising sun"). This Chinese term was also the origin of Polo's "Cimpango" in its various forms. See Lach, Japan in the Eyes of Europe, p. 652, and Lach, China in the Eyes of Europe, 737.

30. NORTH AMERICA

Il disegno del Discoperto della nova Franza, il quale s'ehauuto ultimamente dalle novissima navigatione de' Franzesi in quel luogo . . .

["The draught of the discovery of New France, made from the latest French voyages"]. Bolognini Zaltieri (Paolo Forlani), Venice, circa 1566. From an Italian composite atlas.

Medium : copperplate engraving.
Size of original : 270 x 405 mm.

PLATE : 23

The question as to whether a strait separated North America from Asia was of strategic, rather than just academic, importance, because the existence of such a strait would mean that a northern passage might be found to circumvent the entire continent and provide Europe easier access to Asian markets. This map has roots in both the intellectual and the political facets of the question.

As for the intellectual, Forlani depicts a narrow strait separating America and Asia as had been recently proposed by the Italian geographer Giacomo Gastaldi. Gastaldi had actually long been among the most significant proponents of the theory that North America was connected to Asia; he completely reversed that belief, however, in the last few years of his life. In a map made about 1561 and a pamphlet published in 1562 Gastaldi gave the name *Anian* to a strait which he said separated the Northwest Coast from Asia.[237] The new strait quickly proved to be of enormous influence, becoming almost universally adopted by the end of the century. Judging from the names of both the strait and its adjoining lands, Gastaldi invented his strait from a re-interpretation of part of Marco Polo's text. Polo stated that a large gulf exists which

> *"extends for a two-month's sail toward the north, washing the shores of Manzi on the south-east and of Aniu and Toloman besides many other provinces on the other side."*

The 1561 Gastaldi world map which first showed the strait does in fact place Polo's *Tolman* in the American Northwest (as does, for example, entry 33 in this volume), demonstrating that he believed America to have been the "other side" of which Marco Polo spoke. The term *Anian* almost certainly comes from Polo's *Aniu.*[238]

Forlani's map bears little geographic similarity to Gastaldi's, save for its inclusion of the strait, shown as a very narrow waterway due north of Japan. Below it, the seas between America and Asia are named *Golfo Chinan* ("China Gulf") and *Mare de Mangi* ("Sea of Manzi") after Polo, who said that

> *"the sea in which* [Japan] *lies is called the China Sea —— that is, the sea adjoining Manzi, because in the language of the islanders 'China' means Manzi."*

On the American side of Polo's *Mare de Mangi* lie the political and practical repercussions of the strait question. In 1542, responding to disturbing rumors that the

---

237. The existence of this Gastaldi map was long surmised but it was "lost" until a surviving copy came to light in the mid-1970s. It is a woodcut on nine sheets, and is now in the British Library. Gastaldi's pamphlet, entitled La Universale Descrittione del Monde, was printed by Matteo Pagano.

238. Coolie Verner (The Northpart of America) mentions an obscure (and very unlikely) theory, quoting John Forster who (writing in 1786) speculated that the name was given to Hudson Bay by Gaspar Corte-Real in 1501 in honor of two of his brothers named Anian. According to this theory, Gastaldi adopted the name because Gaspar had believed that the bay might be a sea passage to India.

Portuguese had discovered a route between the Atlantic and Pacific Oceans in the far north, the Spanish Crown instructed Viceroy Mendoza of Mexico to dispatch, without delay, an expedition up the Northwest Coast. Mendoza assigned the Portuguese pilot Juan Rodriguez Cabrillo to the task, and Cabrillo's itinerary became the source of this map's West Coast geography. Beginning from Forlani's *Y di Cedri* (Cedros Island, discovered and named by Ulloa), the map would have Cabrillo skimming Japanese waters on his voyage north to *P. de S. Michel* (San Diego). Continuing up the coast Forlani has marked *P. d. Fuego* after Cabrillo's reported sighting of much smoke inland (signal fires or brush fires). North of it lies *P. de Ogni Sti.* ("all saints point"), which is probably near what is now Los Angeles. Eventually, continuing north, they sighted snowy mountains, the *Sierra Nevada*.[239]

Forlani has also relied on the Spanish expedition of Coronado. Coronado departed Nueva Galicia (northern Mexico) in 1540 determined to reach "Cibola," the fabled Seven Cities which had now flirtatiously retreated to the barely accessible interior of North America. Penetrating the American Southwest, Coronado made his way to the Zuñi village of Hawikuh, shown as *Granata* on Forlani's map. *Civola Hora* above it lies accessible to the *Tontonteac* River, reflecting the Spaniards' hope that the river discovered by Ulloa in the Gulf of California would provide water access to Cibola. The Colorado River itself *(Tigna f.)* flows from Quivera.

In the Northeast, Cartier's second voyage is now recorded with the appearance of the St. Lawrence River and, along its banks, the village of *Ochelaga* (Montreal). Forlani, however, confuses the river *Gamas* of Gomes with the St. Lawrence of Cartier.[240] Cartier's massive inland waterway is labelled *Gamas,* while the name of St. Lawrence *(R. S. Lorezo)* denotes a shorter river flowing south from a large inland lake. Near *La Nova Franza* the name *Canada* now appears, which Forlani himself had introduced six years earlier.[241] Along the Eastern Seaboard are the now-familiar regional names *Laborador, Baccalos, Larcadia, Norumbega,* and *Florida*. The Appalachian Mountains are shown, mis-aligned fully 90°, and the Rocky Mountains are represented in token fashion.

This map by Paolo Forlani was sold without credit by Zaltieri.[242] As a printed map specifically devoted to the North American continent, it is preceded only by the little woodcut in Bordone's *Isolario* of 1528 (entry 18).

- - - - - - - - - - - - - -

239. First recorded by Ramusio ten years earlier, on his "Universale della Parte del Mondo Nuovamente Ritrovata," attributed to Gastaldi. See page 98 in entry 29.

240. Gomes' "Rio de Gamas" ("river of deer") is either the Penobscot River or the Hudson. See page 93 and footnote 221 in entry 27 (Bellero).

241. The first printed map to use the term "Canada" was the 1560 world map of Forlani. It appeared on manuscript maps as early as the "Harleyian" world map of 1542.

242. See David Woodward, The Maps and Prints of Paolo Forlani, 37.01 and 37.02.

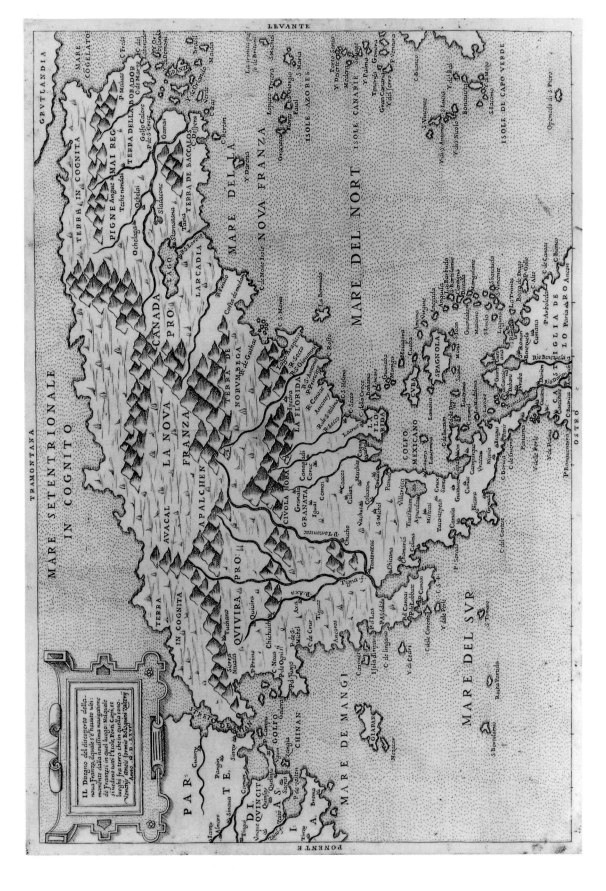

PLATE 23

Paolo Forlani/
Bolognini Zaltieri
Venice, circa 1566

(entry 30)

31. THE WORLD

Orbis Terrae Compendiosa Descriptio Quam ex Magna Universale Gerardi Mercatoris . . .

["Map of the terrestrial world according to Gerhard Mercator"]. Rumold Mercator, Geneva, 1587 (but Amsterdam, 1607).

Medium : copperplate engraving, with original hand color.
Size of original : 290 x 520 mm.

COLOR PLATE : XI

Gerhard Mercator was a student of philosophy; overwhelmingly interested in the origin and nature of the universe, he did not confine his ideas to established concepts. His geography was based on his own interpretation of data (albeit from sometimes dated sources).

Although from a family of modest means, Mercator quickly earned respect in his field. By 1537 he was well-trained in the science of map-making, and by 1541 was successful as a globe-maker. A maverick, in 1544 his inquisitive nature nearly brought him disaster, as he was arrested by the Inquisition for heresy, ultimately spared only because of the intervention of the University of Leuven on his behalf.

His greatest achievement was the construction, in 1569, of a map of the world on twenty-one sheets, the first to use his famous projection; three examples of that work are extant.[243] The present map is an adaptation of it on a double-hemispherical projection, prepared in 1587 by his son Rumold. It first served to accompany a Geneva issue of Strabo's *Geographia*.

Terra Australis, the southern continent or *Antipodes* of classical theorists which had earlier been integrated into Magellan's *Tierra del Fuego,* now appears in mature bloom. It wraps around the fullest expanse of the southern latitudes possible, save for New Guinea, which is correctly shown as a separate island. *Terra Australis* has, in addition, become a dumping ground for errors. It harbors a realm of parrots, *Psittacorum regio* (see entry 19), and serves as the refuge for lands reported by Marco Polo but which have been displaced because they are now duplicated by more recent data. In the East Indies he retains the old *Iava minor* and *Iava maior* of Polo for Sumatra and Java respectively, adds to them a "true" Sumatra, and squeezes between them a promontory of *Terra Australis* labelled *Lucach, Beach,* and *Maleteur. Lucach* is Malaya or Thailand, described by Polo as an independent and gold-rich realm with elephants and other wild animals, *"such a savage place that few people ever go there."* The next region, *Beach,* originated as a copyist's corruption of *Lucach. Maleteur* is Polo's *Malaiur* (Malaya), which Polo said *"conducts flourishing trade, especially in spices."* Mercator's mistaken application of Polo's Southeast Asia to *Terra Australis* results in the displacement of *Iava minor* (Polo's Sumatra) yet further south.

On the opposite pole, Mercator has depicted the Arctic according to a four-island theory previously used by Ruysch and Finaeus. Mercator cited the provenance of this concept in a legend on his 1569 map :

> " [W]e have taken [the Arctic geography] from the 'Itinerium' of Jacobus Cnoyen of the Hague, who makes some citations from the Gesta of Arthur of Britain; however, the greater and most important part he learned from a certain priest at the court of the king of Norway in 1364. He was descended in the fifth generation from those whom Arthur had sent to inhabit these lands, and he related that in the year 1360 a certain Minorite, an Englishman from Oxford, a mathematician, went to those islands; and leaving them, advanced still farther by magic arts and mapped out all and measured them by an astrolabe in practically the subjoined

---

243. A fourth copy was destroyed in the Second World War.

figure, as we have learned from Jacobus. The four canals there pictured he said flow with such current to the inner whirlpool, that if vessels once enter they cannot be driven back by wind." [244]

In the Atlantic between America and Ireland are two islands from Irish myth, *Brasil* (right hemisphere) and *St. Brandam* (left hemisphere).[245] The island of St. Brendan was among the most talked-about islands in the Ocean Sea during the Middle Ages, named for a sea-faring Irish monk and his disciple, Machutus (St. Malo), whom legend states roamed the seas in search of an Isle of the Blessed.

It was Mercator who coined the term *"Atlas"* after the Greek mythical hero. Work on his *Atlas* was slow and deliberate, being finished in stages: in 1578, a Ptolemaic volume was published, which was beautifully engraved but offered nothing of geographic merit, and in 1585 and 1589, volumes covering parts of Europe were published. The third part, embracing the rest of the globe, introducing the name *Atlas,* and containing this world map, did not appear until April of 1595. The old man did not live to see the project finished, having passed away in December of the previous year.

This map was included in editions of the Mercator *Atlas* through 1630 without emendation, thus carrying its fragments of the later Middle Ages into the seventeenth century. Progressively worsening cracking of the plate can be discerned on the map's title. Two letters show cracking in the present example; by the end of the plate's life the left stem of the letter 'A' of *Compendiosa* was affected as well.

- - - - - - - - - - - - - - -

32. AMERICA

Le Nouveau Monde Descouvert et Illustre de Nostre Temps.
/Quarte Partie du Monde.

["The New World Discovered and Illustrated from Our Time
/The Fourth Part of the World"]. André Thevet, Paris, 1575.

Medium : woodcut.
Size of original : 350 x 445 mm.

PLATE : 24

André Thevet was a Franciscan monk and the royal cosmographer to Henry III. He travelled to Brazil both with Guillaume le Testu in 1551 and with Nicolas de Villegagnon in 1555, on the latter expedition settling at *Guanabara,* present-day Rio de Janeiro.

Thevet claims to have based his map of America on his first-hand observations; this is an exaggeration, as much of the map's geography is clearly borrowed from Ortelius. It is nonetheless one of the scarce handful of maps to date whose author had in fact visited America, and it contains considerable information not found on other maps of the period. The region in Brazil where France dabbled in trade and prostelytizing is appropriately called *France Antartique,* as dubbed by Villegagnon, who considered himself to be "King of America." Here Thevet, long anticipating most mapmakers, introduces the name *r Janairo.* Directly inland, Thevet's missionary role in the New World is reflected in his knowledge of various Indian peoples. Most prominent is *Toupinambaux,* the Tupinamba Indians known to us mainly through the adventures of the German adventurer Hans Staden, whom they had held

---

244. Translation from Fite & Freeman, A Book of Old Maps.
245. See page 34 in entry 9 (Beccari) for Brasil Island.

captive shortly before Thevet's second trip.[246] The map's peculiar depiction of southwest South America is unique, and forms an interesting bridge between maps showing a bulge in that region and subsequent maps which corrected it.[247]

Along the Eastern Seaboard of North America, which he skirted before crossing the Atlantic on his return to France in 1556, Thevet adds some nomenclature to his basic Ortelius prototype. A "Sandy Cape" *(C. Sande)* is probably the *Cabo de Arenas* of Gomes, here appearing as the southern tip of Nova Scotia (Cape Sable) but positioned precisely on the latitude of Cape Cod (to which feature *Cabo de Arenas* is in fact sometimes applied) because of the erroneous lateral orientation of the coast. In the waters off eastern Canada Thevet revives Thule, Ptolemy's Shetland Islands, already forgotten by many mapmakers. His use of extraordinary nomenclature and occasionally odd geography persists throughout the rest of the map, in regions to which he had never been and therefore based on sources other than his own experience. In the Northwest Coast, for example, he introduces the rare (and unlikely) French variant of *Montagnes negres* ("Black Mountains") for the Sierra Nevada, *negre* probably being a corruption of *neve* (snow). In the Pacific Ocean *Cimpegu*, the Japan of Marco Polo, denotes a small island off the west coast of New Guinea. Thevet flippantly depicts *Terra Australis* thriving with sub-tropical life in a biosphere which continues unchanged through New Guinea, and leaves the question of New Guinea's relationship to *Terra Australis* open, using his address to the reader to cover the strategic area and avoid the issue.

The quality of Thevet's woodblock is extremely sophisticated, reflecting the peculiarly French finesse for fine woodcuts. At a time when the copperplate had already become the preferred medium, this work's detail and refinement is vastly superior to the woodblock maps still being produced in Germany, Switzerland, and to an extent even in Italy. Thevet was a strong proponent of the theory that the indigenous people of America were the descendants of the lost tribe of Israel; his enthusiasm for the idea was instrumental in propelling it into the next century. The subtitle of his map of America, "the fourth part of the world," is a carry-over from Waldseemüller and his 1507 treatise about Vespucci.[248]

- - - - - - - - - - - - - -

---

246. Thevet and Staden appear to have witnessed similar rituals in Tupinamba life; e.g., both offer a similar description of a Tupinamba burial (unless, as is possible, the apparent corroboration is an illusion resulting from plagerism on the part of Thevet).

247. The bulge in South America was popularized by Mercator in 1569 and Ortelius in 1570. See entry 31 (Mercator).

248. See page 54 in entry 13 (Apianus).

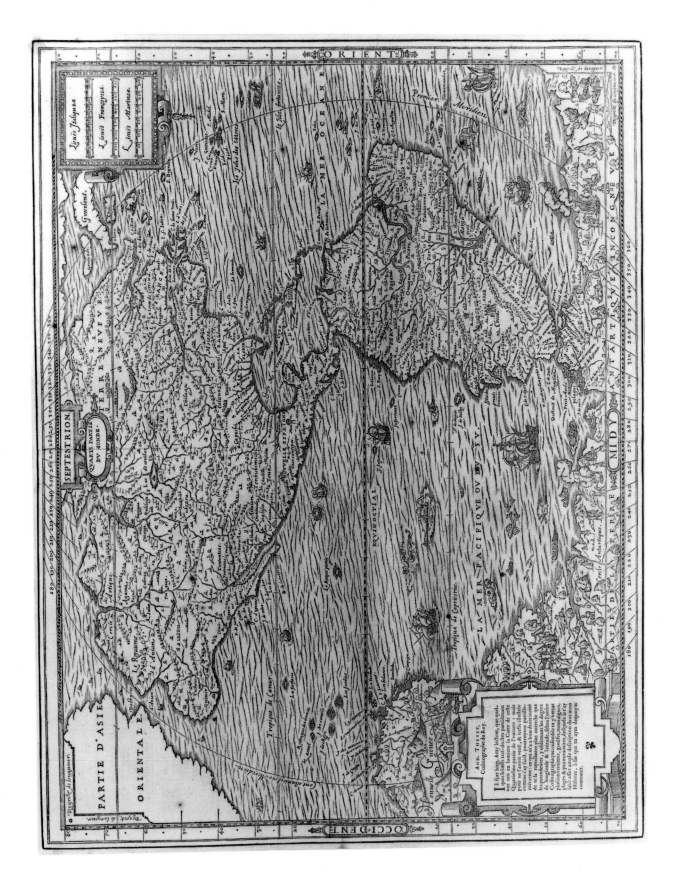

PLATE 24

André Thevet
Paris, 1575

(entry 32)

33. GLOBE

[untitled terrestrial globe].

Imprint reads *Marius Cartarus Viterbiensis autor incidebat Romae MDLXXVII cum privilegio.* Mario Cartaro, Rome, 1577.

Medium : copperplate engraved gores mounted to a partially hollowed wooden sphere, with original hand color.
Size of original : 490 mm circumference; 160 mm diameter.

COLOR PLATES : XII and XIII

The earliest surviving terrestrial globe dates from the twilight of the pre-Columbian period, made from manuscript vellum gores pasted to a plaster sphere by Martin Behaim in 1492. Behaim filled in the as yet unknown reaches of the earth with islands thought to lie in the Ocean Sea; his view of the earth coincided closely with that of Columbus. By the first decade of the new century, the revolutionary discoveries had given cartographers new shores to fill the "far" side of their globes, and the medium of printing had been tapped to reproduce globes in quantity.[249] Globes became coveted intellectual objects of art, ranging in size from portable pocket versions to large floor-standing productions reserved for the elite few. Desk globes, small enough to sit on a table but sufficiently large to record detailed geographic data, became the centerpiece of many a nobleman's study. By their nature, globes were exceptionally vulnerable and extremely few have survived.

This desk globe by the Italian maker Mario Cartaro is made from twelve engraved paper gores mounted on a partially hollowed wooden sphere. The engraving quality is typical of the finest Italian makers, for whom elegance and clarity were overriding concerns, and superfluous decoration was avoided. The globe was produced early in Cartaro's career. A native of Viterbo, Cartaro moved south first to Rome, where he established a fine reputation as an artist/engraver, and by 1582 to Naples, where he worked for the remainder of his life. It was during his time in Rome that he produced this terrestrial globe and a companion globe of the heavens.

The known globes of Cartaro are all dated 1577, although he made at least two different terrestrial models. The present specimen records more advanced geographical knowledge than the other recorded Cartaro globes, following generally (though not exclusively) the Gastaldi world map of circa 1561.[250] North America extends north to well beyond the Arctic Circle, and contains two large inland lakes. One, situated towards the southeast, is unnamed but correponds to the *Lago Paga* in the Gastaldi map. The other is marked *Cani*(bas), the enigmatic Lake Conibas which on Cartaro's globe lies in a desert named *Zubican*.[251] In the Northeast is *Picnmai R* (i.e., "Picne Maay Regio"), a kingdom of one-legged pygmies which Cartier claimed to have learned about from his Indian informants on his second voyage. This story, however, is suspiciously reminiscent of the *Skrealing* and *Ein-foetingr* legends of early Norse visits to Greenland.[252] North America is separated from Asia by Gastaldi's Strait of

---

249. The Waldseemüller gores of 1507 were printed from a woodblock, although no mounted specimen is known to survive. The earliest extant post-Columbian globe is the anonymous "Lenox" globe of circa 1510 (engraved copper sphere).

250. See illustrations of Cartaro globes in Stevenson <u>Globes</u>, Volume I, facing p. 168, and Shirley, <u>Mapping of the World</u>, plate 116 (unmounted gores).

251. See page 125 in entry 41 (de Jode) for Conibas Lake.

252. A land of pygmies was in fact placed in pre-Cartier "Amerasia"; see page 75 and footnote 192 in entry 20 (Finaeus).

Anian (unnamed) with the Asian kingdoms of *Tolman* and *Agama*[253] situated on the American Northwest, and the Asian kingdom of *Anian* occupying the Asian Northeast, conforming to Gastaldi's interpretation of Marco Polo.[254] Seventy years earlier Ruysch, interpreting the same text, placed *Tolman* near Cathay and Tibet, and understood *Agama* to be an island due west of Sumatra.

The term *America* had remained a surprisingly unpopular commodity on Italian maps for much of the sixteenth century. Perhaps it was a victim of city-state rivalry; Rome and Venice, rather than Vespucci's Florence, were the major centers of map production. By the mid-sixteenth century, however, the name appeared on Italian maps derived from other (non-Italian) works, and notably on the non-derivitive Tramezzino hemispheres of 1554. Cartaro also accepts the term, but like Tramezzino he limits its usage to South America.

Cartaro's Southern Continent, named only as *Terra Incognita,* reaches into the Southwest Pacific to north of the Tropic of Capricorn. He does not, however, follow Mercator and Gastaldi in erroneously placing Polo's Malay nomenclature there, and deviates from Gastaldi in adding New Guinea, shown correctly as an island. In the Indian Ocean the northern segment of the southern continent is marked *Terra di Vista* ("land which has been seen"). Portuguese advances in the Orient are evident in Cartaro's clearer picture of Southeast Asia, Thailand now appearing by its "modern" name *(Siani).* In the Indies both Spanish and Portuguese data has enabled a diminishing reliance on Marco Polo, with Borneo *(Burnei),* first known from Pigafetta, and the Indonesian islands being recognizably shown.

- - - - - - - - - - - - - -

34. SOUTHWEST SPAIN [Sea Chart]

Zee Custen van Anduluzien . . .

["The sea coast of Andalusia . . ."] [in *Spieghel der zeevaerdt . . .*
Lucas Janszoon Waghenaer, Leyden, 1583 (1586).]

Medium : copperplate engraving, with original hand color.
Size of original : 325 x 505 mm.

COLOR PLATE : XIV

In response to an escalating demand for navigational charts, the first printed *rutter,* or mariner's guide with a full set of sea-charts, was introduced at the end of the sixteenth century. It was the inspiration of a Dutch pilot, Lucas Janszoon Waghenaer, who received privileges for it in 1579 and 1580, and published the first part of it in 1583.

Waghenaer grew up in the port of Enkhuizen in the West Frisian region of Holland. Although the trade in herring from the North and Zuider Sea fisheries had provided him as a young man with a brisk industry, the unyielding yoke of Spanish control stifled any incentive for him and his countrymen to pursue their own commercial designs. As a result, some Dutch entrepeneurs looked to trade with parts beyond Spain's control, specifically northern Europe

---

253. See entry 12 (Ruych) for an example of both Tolman and Agama mapped as Asian lands. In that work Tolman is part of China and Agama is an island; Agama is possibly a corruption of Polo's "Dagroian," a kingdom on Sumatra (but not the kingdom in which he spent five months) whose people Polo said ate the entirety of its dead because otherwise worms would generate from the body, then die of hunger and torment the soul of the deceased.

254. See page 101 in entry 30 (Forlani /Zalteri).

and the Baltic region. Waghenaer himself probably accompanied a trading voyage to Lapland in 1565.

When Waghenaer's region of Holland eventually freed itself from Spain, the urge to trade freely overwhelmed all constraints. Lucrative trading, as well as smuggling and even piracy, ensued on a licentious scale. Their bounty was especially rich because Amsterdam, traditionally a competitor, was still under Spanish rule. West Frisia's need for sea-charts was stronger than ever, and Waghenaer, well-respected as a pilot, began compiling his charts as early as 1570.

While working on the project, Waghenaer earned his livelihood as Enkhuizen's tariff collector. But in 1582, well into the extremely costly preparation of the *Spieghel,* he was dismissed from the post for improprieties. He then endured the financial hardship of pursuing the atlas and supporting his family on earnings from meager jobs. Fortunately, his atlas was successful. Its lusciously produced plates rivalled the elegance of its manuscript forebears and established a new genre of the printed medium.

This enlarged 1586 issue of Waghenaer's *Spieghel* is shown opened to the map of Andalusia, Spain. On the far left is the *Rio Tinto,* down which Columbus sailed from Palos and into the Atlantic on August 3, 1492. Just to the west of the *Tinto* and sharing a common delta is the *Rio Saltés,* the river by which the last exodus of Jews fled the Inquisition. By coincidence, August 2 had been the deadline mandated by Ferdinand and Isabella for all Jews refusing conversion to vacate the country or be executed, and so that last shipload of refugees left on the same tide as Columbus.

After the success of his *Spieghel der Zeevaerdt* Waghenaer produced another sea-atlas, the *Thresoor der Zeevaerdt,* on a smaller oblong format. That less luxurious but perhaps more practical format was then copied by the emerging Amsterdam maker, Willem Blaeu, in 1608. The sea-chart, or printed portolan chart, had taken firm root in Holland.

- - - - - - - - - - - - - -

35. PACIFIC OCEAN AND AMERICA

Maris Pacifici, (Quod Vulgo Mar del Zur) . . .

["The Pacific Sea, (commonly known as the South Sea) . . ."].
Abraham Ortelius, Antwerp, 1589. From the *Theatrum Orbis Terrarum.*

Medium : copperplate engraving, with original hand color.
Size of original : 345 x 495 mm.

COLOR PLATE : XV

The editions of Ptolemy's *Geographia* published in Bologna and Rome in 1477 and 1478 respectively were the first printed atlases. These were based strictly on classical knowledge and were necessarily limited in geographic scope. During the following several decades, new lands did quickly make their way into atlas maps, though the atlases themselves remained classically oriented. By the middle of the sixteenth century a "composite" atlas could be made-to-order from the stock of Italian mapsellers; these atlases, though finally shedding classical bonds, were simply collections of otherwise unrelated loose-sheet maps. The first true atlas in the modern sense of the term, being a set of uniformly produced and methodically organized maps of all parts of the world based strictly on current knowledge, was the *Theatrum* of Abraham Ortelius, first published in 1570.[255] The atlas was commercially successful and continued to be updated and enlarged through its final editions of 1612, outliving Ortelius himself, who died in 1598.

From the beginning of the *Theatrum's* life, Ortelius' concern for commercial viability was evident in the beauty of its maps. Most of the atlas' significant maps were not present in its original form, however, but were added during the course of its forty-two year evolution, demonstrating that Ortelius also possessed the courage and initiative to use fresh data to chart regions which were still poorly understood. A prime example of this is his map of the Pacific, which was engraved in 1589 and first included in the atlas in 1590. This is the first printed map specifically devoted to the Pacific Ocean. It covers the half of the earth lying on Spain's side of the *raya,* and may indeed owe its inception to a Spanish chart. The proportion of the globe encompassed by its 180° of longitude is fairly accurate, however, exhibiting none of the self-serving manipulation of the line of demarcation shown by some Spanish charts of the period. The source of Ortelius' data remains an enigma.

As compared to his maps of just two years prior, he has radically revised the northwest coast of America, the California peninsula, and the contour of the west coast of South America. Ortelius left the land areas almost completely denuded of interior detail, reflecting either his intended emphasis on the ocean itself, or perhaps his reliance on a portolan chart prototype. One of the few inland features, Mexico City, is now more accurately located. The mythical Seven Cities sought by Coronado have mercifully vanished, although *Quivera* still occupies the far Northwest. And finally we see the distinction made between "North" and "South" America, the continents being designated *Americae Septemtrionalior Pars* and *Americae Meridionalior Pars*.[256]

The name *Califormia* (sic) now appears, though designating only the peninsular region. Use of the term can be traced back as far as 1541.[257] It appears to have been coined after an imaginary place in a Spanish sequel to an originally Portuguese romance novel entitled *Amadis de Gaula.* The author of the sequel, Garcia Montalvo, describes *"an island called California, very near to the region of the Terrestrial Paradise,"* which is to say *"on the right hand of the Indies,"* where indeed early explorers believed California to lie. Montalvo's *California* was an island inhabited solely by black women, a slight variation on the general Amazonian theme of eager, if dangerous, women, a theme which clearly occupied many a sailor's fancies.[258]

Along the Pacific coast of the peninsula, a point marked *Cabo del Engaño* suggests a Spanish prototype. This "Cape of Deceit" was so-named by Ulloa because of his frustrated attempts to find a passage above what he presumed to be the *island* of California. Other nomenclature, such as the Italianized *Messico* and the appearance of duplicate place-names (of the *Bravo* and *Grande* Rivers), suggest that two or more prototypes were combined.

On Ortelius' previous maps, New Guinea had been variously charted as an island or part of *Terra Australis,* and accompanied by a disclaimer stating that its nature was unknown.[259]

---

255. The term "Atlas" itself was coined by Gerhard Mercator, and first used, though posthumously, in 1595.

256. The term "America" was first used in 1507 (see entry 13, Apianus), limited only to South America; it was applied to both continents as early as 1513-15 on the "Paris Green" or "Quirini" globe, and by Mercator in 1538 (see entry 22, Florianus), without distinction made between the "northern" and "southern" part.

257. The term "California" is found on the Domingo del Castillo map of 1541 (manuscript). See Wagner, Northwest Coast, p. 32.

258. See Polk, The Island of California, p. 124, 130.

259. Although New Guinea had been discovered possibly as early as 1511, the question of whether it was an island or part of "Terra Australis" remained unanswered until Torres' incredible voyage of 1605. But as his discovery was concealed for secrecy and officially forgotten, the question of its relationship to Australia remained until the first voyage of James Cook.

But now he correctly shows it as an island, modified in shape (though no more accurate), and despite focusing on the problem more directly than the previous maps, there is no longer a legend qualifying his choice of insularity. *Something* had made him considerably more confident of its insularity in the two years that had elapsed since 1587. Off the east coast of New Guinea now lie the Solomon Islands, discovered by Alvaro de Mendaña during his voyage from Peru in 1567-69. These islands, whose namesake evoked images of colossal riches, first appeared on Ortelius' maps of 1587 in a somewhat different configuration.

Although much can be said about Ortelius' new data on the continental shores of the ocean, perhaps the most revealing feature of the Pacific itself is its utter absence of major archipelagos. Most of the principal island groups of the Pacific remained invisible to European vessels until the later eighteenth century, making early trans-Pacific travel an ordeal of stamina. Knowing nothing of the lush Polynesian islands which would marvel Europe in two centuries, Ortelius has adorned the unknown expanse of the South Pacific with an illustration of Magellan's ship, the *Vitoria*.

- - - - - - - - - - - - - -

36. AMERICA [Wall Map]

Nova et Acurata Totius Americae Tabula au G.[B.].

["A New and Accurate map of all of America by G. Blaeu"]. Willem Blaeu, Amsterdam, circa 1608 (Pietro Todeschi, Bologna, circa 1673). Separate publication.

Medium : copperplate engraving, with original hand color.
Size of original : 1,040 x 1,420 mm.

COLOR PLATE : XVI

Antwerp, which had been the center of such prolific map makers as Ortelius, was sacked by the Spanish in 1576 and eventually fell to them in 1585 after a 14-month siege. In retaliation the Dutch blocked sea access to Antwerp via the Scheldt River, strangling the city's commerce. The tactic inevitably wreaked havoc not only for the Spanish, but for Dutch interests in Antwerp as well. A prime beneficiary was one of Antwerp's rival Netherland cities, Amsterdam.

By the turn of the seventeenth century, cartographers in Amsterdam had secured pre-eminence in the publishing of maps, including the production of spectacular large works intended to adorn the walls of wealthy patrons. Willem Blaeu, from early in his career, had staked much of his commercial success on these works. His wall maps became highly prized and as a result were plagiarized by other makers. The present map is a copy of Blaeu's wall map of America, originally published in Amsterdam circa 1608. Although no copy of the 1608 issue is known to survive, the map was periodically resurrected by other makers during the century and is here copied by the Bologna publisher Pietro Todeschi circa 1673. Despite its Italian provenance, this work epitomizes Dutch cartography at its zenith.

Dutch and Flemish map-making, the pre-eminence of which began about 1570 with the introduction of Ortelius' *Theatrum* and lasted through the close of the seventeenth century, was closely integrated with Dutch art in general and coincided with the lives of many of the great Dutch artists. Indeed, many map-makers were influenced by their artist compatriots, just as some artists incorporated the map as art object into their canvases. Most familiar among these was Jan Vermeer, whose *Lady with a Lute* and *Artist in His Studio,* among others, contain hanging maps as compositional elements, demonstrating the popularity of the wall map as fashionable adornment for the walls of the affluent.[260]

Wall maps were envied well before the Dutch period, however. The use of such maps to dress-up the dwellings of the privileged can be traced back to at least the later Middle Ages, when magnificent and quasi-encyclopedic *mappaemundi* were created for the rich, the State, and the Church.[261] With the advent of printing, the woodcut medium was used to produce large city-views as early as 1500, and elaborate, monumental maps of the world at least as early as 1507.[262] The wall map's transition to the superior copperplate medium occurred at the hands of the Italian school in the second half of the sixteenth century.

Geographically, the map utilized the finest data available to Blaeu in the first decade of the seventeenth century. But by the time Todeschi issued this copy in the third quarter of the century, much of it was dated. Even the map's depiction of California, correctly shown as a peninsula while virtually all contemporary maps adhered to the island theory then in vogue, should fairly be judged as an anachronism rather than as a virtue. The only obvious benefit Todeschi reaped from the intervening years appears in the inset of the south polar region, where the discoveries of Abel Tasman in 1642 are recorded (mistakenly inscribed *detecta 1667*). But because the discovery of Australia in 1612 is not shown, Todeschi's Tasmania lies among the remnants of *Boach Provicia* (a mis-charting of Marco Polo's Southeast Asia) and other vestiges of the defunct Southern Continent. The discovery by Le Maire of a route around Tierra del Fuego in 1615-16 is noted both on the inset and on the map proper but Blaeu's pre-Le Maire geography has been left essentially intact.

Just as Todeschi plagerized Blaeu's map, the illustrative panels used by Blaeu were themselves pirated from other works. The panels along the left and right sides, illustrating the customs and dress of the various peoples of America, are derived from the *Voyages* of Theodore de Bry.[263] Of these, six along the left border pertain to North America. The second from the top shows the manner in which the women of *Dasemonquepeuc* (Virginia) carry their children, a young woman of *Secota* (Virginia), and chief lady of *Secota*; third from the top shows a chief lady of *Pomeiooc* (Virginia) carrying a gourd containing "some kind of pleasant liquor," and an old man of *Pomeiooc* in winter clothes; fourth from the bottom shows a religious man of *Secota,* a conjurer or magician, and a prince set for battle; fifth down depicts a chief and warriors of Florida; sixth down, also Florida, is a king and his first wife out for an evening stroll in the forest, assisted by an attendant; the next depicts sovereigns of Hispaniola; and the second from the bottom shows a king, queen, and attendants in *Nova Albion,* Drake's landing on the West Coast generally believed to be California. Some of the town views flanking the bottom are taken from the Portuguese cartographer Baptista Boazio.[264]

Like globes, wall maps suffered the ravages of time and the elements to a far greater degree than did maps bound into atlases or other books, and few have survived. Compounding their vulnerability, they were often coated with varnish which yellowed and cracked with age. Vermeer's painting testifies to the rapidity with which such decay began; a degree of deterioration can be noted even on the Visscher wall map of the Netherlands hanging in Vermeer's *Artist in His Studio,* frozen in time by Vermeer as it fared in about 1667.

- - - - - - - - - - - - - - -

260. See James A. Welu, "The Map in Vermeer's Art of Painting," Imago Mundi 30 (1978).

261. See entry 2 (Le Rouge) for a map with geography typical of this genre.

262. E.g., the six-sheet view of Venice by de'Barbari, Venice, 1500, and the two large world maps of Martin Waldseemüller, Strassburg, 1507 and 1516.

263. Frankfurt, 1590, and subsequent editions.

264. Boazio made his engravings from plans made of Francis Drake's attacks on West Indians strongholds in 1585-86.

37. THE WORLD

Nova Delineatio Totius Orbis Terrarum Auctore A. Colom.

["A new rendering of the entire terrestrial world"]. [in: *Atlas marin, Ou Monde Maritime. Contenant une briefe Description de toute les costes coqnues de la Terre. Nouvellement mis en lumiere, par Arnold Colom . . . Amsterdam, A. Colom . . .*] Arnold Colom, Amsterdam, circa 1655.

Medium : copperplate engraving, with original hand color.
Size of original : 555 x 625 mm.

COLOR PLATE : XVII

This map of the world is an unusually bold presentation of a classic esthetic ideal, exemplifying Baroque cartographic art. In it Arnold Colom's competent (though otherwise undistinguished) geography is supported by six allegorical figures, being the four elements and Day and Night. Fire is at the upper left, Air is at the upper right, Earth at the lower left, and Water at the lower right. At the top and bottom of center are Day and Night.

In North America, the "true" Great Lakes, rather than the possible allusions to them found in the *Mare Dulce* and *Conibas Lake* of the previous century, appear. They are however recorded only as a southerly extension of Hudson Bay, in which respect Colom's map is far behind the finest contemporary French maps. The Northwest Coast, dubbed *Nova Albion* after Drake's North American landfall, is depicted as a huge continental shoulder bridging much of the North Pacific.

Colom records Le Maire's strait around *Tierra del Fuego* into the Pacific, although like most of his contemporaries he has not altogether abandoned the belief that *Terra Australis* might exist. Australia has been added to the world picture based on Dutch excursions from the Indian Ocean and New Guinea. The reports of Abel Tasman's circumnavigation of Australia (1642-44) have not, however, been tapped, except to acknowledge that the land's western and southern shores are indeed independent of *Terra Australis*. Neither Van Diemen's Land (Tasmania) nor New Zealand, both known from Tasman, yet appear.

Although the continental shores which define the Pacific Ocean have evolved during the two-thirds of a century since Ortelius' *Maris Pacifici* (entry 35), the great ocean itself has changed little. Some new islands have been added to its repertoire, but its size has remained nearly stagnant. As a measuring reference, consider the longitudinal breadth between America and Japan. The following comparison is based on the distance between the east coast of Japan and the westernmost point of equatorial South America, present-day Equador/Peru (the west coast of North America is not used here as a reference, even though it is visually more convenient, because that still-unkown coast itself shifted about and thus introduces another variable into the equation) :

In the pre-"discovery" period of Japan, such as the 1507 world map of Waldseemuller (entry 13, plate 9), about 20° typically separated the two worlds in an essentially conceptual Pacific. A transitionary step, after Magellan conquered the ocean and revealing post-Polean knowledge of Japan, is found in the 1556 hemisphere of Gastaldi (entry 29, plate 22). In that work the ocean between Japan and South America spans about 85°. In 1589 Ortelius (*Maris Pacifici,* entry 35, plate XV) judged approximately 120° to separate Peru from his now post-contact Japan; here on Colom's map in the mid-seventeenth century, the figure has remained virtually unchanged, but a foreshortening of the west coast of North America and a more accurate placement of New Guinea further to the west have "opened-up" the ocean. The true stretch of ocean separating Japan from South America is about 145°.

- - - - - - - - - - - - - -

PART THREE — EARLY COLONIZATION

The first post-Columbian European colony in America was settled by mishap rather than by design when Columbus, in order to accomodate the surplus of sailors caused by the loss of the Santa Maria *to a coral reef at the end of 1492, founded the village of* La Navidad *on Hispaniola. La Navidad, like most early colonies, was doomed from its beginning and ended in disaster. Although attempts at colonization continued to be perplexingly poorly planned, maps did quickly assume an increasingly vital role in all aspects of colonization. They enabled settlers to choose more suitable sites, allowed cargo ships to make ports-of-call to supply them, helped colonists locate sources of food and objects of trade, served as advertisements to recruit new settlers, helped to plan strategy for their defense, graphically brought security concerns to the attention of their mother countries, and were consulted as documents to legitimize claimed boundaries. Pictorial elements on maps also helped settlers recognize that America was culturally, as well as geographically, a truly new world.*

Colonization, in turn, swiftly altered the face of the map itself. Such was the aggressiveness of European relocation to America that claims were mapped over vast expanses of the continent by the early seventeenth century, reserving land in anticipation of the hordes yet to come. This mentality of "manifest destiny" was bluntly illustrated by a British settler in 1670 : "it hath generally been observed," *he wrote,*

> "that where the English come to settle, a Divine Hand makes way for them, by removing or cutting off the Indians, either by Wars one with the other, or by some raging mortal Disease." [265]

The following maps figured into the deliberate, methodical attempts at colonization which established Europe's irrevocable grip on American soil and her eventual dominance over the continent.

---

265. Daniel Denton, <u>A Brief Description of New York</u>, London, 1670, p. 7.

Chapter VII

THE BRITISH AND FRENCH

38. THE WORLD

La Heroike Enterprinse Faict Par le Signeur Draeck d'Avoir Cirquit Toute la Terre.

["The heroic enterprise of Sir Drake, who has circumnavigated the world"]. Nicola van Sype, Antwerp (?), 1583. Separate publication.

Medium : copperplate engraving.
Size of original : 240 x 440 mm.

PLATE : 25

Sir Francis Drake's circumnavigation of the world (1577-80) can be seen as the beginning of the end of Iberian domination in the rush to lay claim on the non-European world. Indeed, the voyage was a flamboyant display of England's determination, competence, and brawn, with which the other European powers would have to reckon.

The voyage was also proof of political savvy: when Drake left Plymouth in 1577 his crew, and in fact most of the British government itself, was entirely ignorant of their true mission, believing instead that they were headed for a routine trading voyage to Alexandria and the Levant. The expedition was truly covert. Only once well at sea did the crew learn the true nature of their daring and ruthless mission.

As plotted on van Sype's map, Drake attacked Spanish strongholds in South America, and when his ship was laden with plunder and damaged from battle, he sought refuge far north to repair and restock, reaching a bay at *Nova Albion* on the west coast of North America. The identification of this anchorage has aroused fervent dispute among modern historians. Van Sype himself appears to have been unsure of the location of the bay: he plots conflicting lines, one recording a landfall considerably north of the other.[266] In this regard it is interesting to note the inscription on the bottom of the map stating that it has been seen and "corrected" by Drake himself. It is possible that the precise location of Drake's landfall was in fact deliberately obscured by the British government to conceal knowledge of an inlet or river which they hoped might lead to a Northwest Passage; it is perhaps even more likely that the British wanted the Spanish to *believe* that they had discovered such a passage.[267]

Upon leaving the West Coast of North America, Drake crossed the Pacific without sighting any land, eventually reaching shores he called *Island of Thieves* in response to its people's disinterest in the concept of private property. As with Drake's North American landfall, the identification of this island has been contested by historians. It is variously thought to have been one of the islands in the north-south oriented Mariana chain, or one of the Carolines, or Palau, or one of the Philippines. Similar experiences of islanders' stealing habits had led Drake's only predecessor in global circumnavigation, Magellan, to dub his first

---

266. The same ambiguity is found on the Hondius "Drake" map of circa 1595 and on the "Drake Mellon" map (manuscript, after 1586).

267. See page 143 in entry 48 (de Fer).

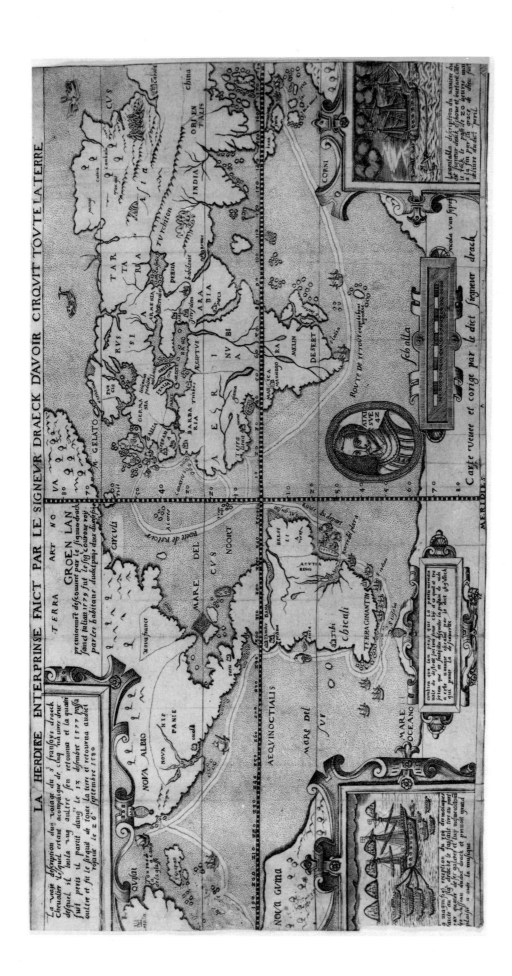

PLATE 25   Nicola van Sype, Antwerp (?), 1583   (entry 38)

landfall in the western Pacific with the same "Island of Thieves" name *(Ladrones)*.[268] Van Sype records Drake's landfall at its approximately correct latitude between the equator and Tropic of Cancer, but labels it only as the *Moluccas,* at that time a somewhat generic name for any of the "spice islands" lying to the south.

Insets on the map depict Drake being received by the Sultan at Ternate, and one of his ships hitting rocks off Celebes. Elizabethan coats-of-arms appear at his landfall on the California(?) coast, and at the southern extreme of South America. British claims in southernmost America were strategic, for it was Drake who is first known to have suspected that open ocean lay close at hand south of Tierra del Fuego. For a maritime nation, this would have meant a promising new route into the Pacific. Drake's theory of an insular Tierra del Fuego was flaunted about intellectual circles and was ultimately tested by Dutch, rather than English, interests. In 1615 a consortium of Dutch merchants sent Jacob Le Maire in search of the strait which now bears his name; by proving Drake correct, Le Maire both revealed an alternate route between the Atlantic and Pacific Oceans and, more importantly to his backers, circumvented the Dutch East India Company's monopoly on trade through the Magellan Strait.[269]

In October of 1580, Drake presented Queen Elizabeth a map recording his circumnavigation. Van Sype's map is probably a copy of that map, which in 1625 Samuel Purchas described as *"still hanging in His Majesties Gallerie at Whitehall, neere the Privie Chamber."*[270] This map of van Sype is the earliest printed work to show Drake's voyage. Having served as a broadside to announce the news of Drake's circumnavigation and his exploits, it is a somewhat carelessly executed production, suggesting the author's rush to beat others to the press with fresh and still exciting news. Its estimated date of 1583 is based mainly on the map's portrait of Drake as 42 years of age, which would place the map at this time, his birth thought to have been in 1540-41. It was presumably done before news was received of Drake's West Indian raids (1585-86), the tracks of which are not shown.[271]

Sir Francis Drake's brilliant military successes against Spanish targets in America signalled the decline of Iberian pre-eminence in Europe's overseas imperialism. With them, England had proven herself a superpower ripe to attempt her own colonization of the New World, while both Spain and Portugal (who for six decades, beginning in 1580, existed under a united crown) had fallen into disastrous debt in their zeal to extend their spheres of influence. Drake himself helped inflict final, decisive blows against Spain both during his West Indian attacks of 1585-86 and the defeat of the Spanish Armada in 1588. Portugal's overseas enterprises fared comparably : England and Holland shared the spoils of Portugal's decaying empire in the Orient, superseding her as the dominant forces in India by the early seventeenth century.

- - - - - - - - - - - - - -

268. This was in 1521. Magellan had originally dubbed it "Island of the Lateen Sails" because of the sailing vessels found there, but changed the name when stealing become a problem. His landfall is commonly believed to have been Guam in the Marianas.

269. But Le Maire, upon reaching his country's outpost in Batavia, was incarcerated by the local Dutch officials for infringing on the Dutch East India Company's monopoly. They did not believe, or did not wish to believe, that he had entered the Pacific by an alternate route. Le Maire was sent back to Holland in chains and died en route.

270. Quote from Samuel Purchas, Purchas his Pilgrimes, London, 1625, part III.

271. The van Sype map closely resembles the "Drake-Mellon" manuscript map which is believed to be copied from the Queen Elizabeth map except that it includes the tracks of Drake's West Indian voyage of 1585-86. The map of North America by Michael Lok, probably done a year earlier than van Sype's (1582), makes simple verbal reference to Drake's presence on the California coast ("Anglorum - 1580").

39. MID-ATLANTIC COAST

Americae pars, Nunc Virginia dicta . . .

["The part of America which is now Virginia . . ."].
John White, Frankfurt, 1590. From Part I of de Bry's *Voyages*.

Medium : copperplate engraving, with original hand color.
Size of original : 305 x 420 mm.

COLOR PLATE : XVIII

English precedent in North America can be traced back to at least 1497, when the Italian explorer John Cabot sailed to northern American coasts under the British flag. Richard Hakluyt, a fanatical proponent of British expansion, claimed an even earlier British precedent by citing the twelfth-century discovery of America by the mythical Prince Madoc.[272]

The mid-Atlantic coast was the clear choice for England to make her foothold in North America for several reasons. It was wise to look south of the Cabot and purported Madoc landfalls, both because France had already established claims there, and because that region's coastal waters were notoriously treacherous. To the south, however, she had to keep clear of current Iberian interests in Florida. Further, it was still hoped that a sea route through the New World might be found along the mid-Atlantic coast where Verrazano had reported to have seen the China Sea, and control of such a passage would be enormously profitable. Lastly, Sir Humphrey Gilbert, under letters patent from Queen Elizabeth, had sent a Portuguese pilot named Simao Fernandes on a preliminary scouting of the Atlantic coast (1580), and the chart Fernandes prepared from his survey showed a *Bahia de Santa Maria,* placed at the latitude of Pamlico Sound, as the only obviously beneficial site.

In 1584, one Walter Raleigh, at that time noted for his valor in fighting Irish rebels, was granted a patent for an expedition to North America. The voyage was undertaken that year, though the Queen forbade Raleigh himself to go. Raleigh's Elizabethan envoys reached and penetrated the Outer Banks that had confused Verrazano sixty years earlier and, like Verrazano, they never reached the mainland. Instead their fancy was caught by an island which was called *Roanoke* by the Indians. The character of Roanoke Island is obvious from White's map: it was amply shielded from the ocean by the Outer Banks and lay in a pivotal position between Pamlico and Abermarle Sounds. Access was restricted. It could be reached only by shallow inlets through the Banks, which though inconvenient offered much as regards protection from foreign predators. The local Indians, who were the most southerly of the Algonquian tribes, were scrupulously friendly and catered to their British visitors' needs. Roanoke was also safely distant from the village of *Neuustooc* (far left) on the River Neuse, said to be inhabited by "hostile" Indians (probably Siouan-speaking Catawbas).

Back home the following year amid lavish accounts of the land's virtues, the "Virgin Queen" knighted Raleigh and immodestly allowed their new soil to be called "Virginia" in her honor. Plans were quickly made for a permanent settlement. At Hakluyt's recommendation, John White, an excellent artist and draughtsman, was asked to join the expedition.

The fleet approached North America by way of the Caribbean, reaching the mainland south of Cape Fear (White's *Promontorum tremendum*). They followed the coast north to an inlet marked *Wokokon* which, as White has indicated, has a large area of shoals. Some of the fleet's seven ships ran aground on them. While crews were busy refloating them, Richard Grenville, commander of the enterprise, led a scouting party in smaller boats into Pamilco Sound and River. After being well received at the town of *Secota* (Secotan), metropolis of the Secotan Algonquian Indians, they pillaged another village in response to the loss of a

---

272. Or possibly semi-mythical. Madoc ap Owain Gwynedd (circa 1170) was said to be a Welsh prince, son of Owain Gwynedd. The purported discovery of Welsh-speaking Indians has been cited to support the story.

silver cup. The full fleet then continued north to *Hatorask*,[273] at which point a hundred settlers, under Ralph Lane, penetrated the Banks and claimed the island of *Roanoac* as their new home. They built a fort on its north end, facing Albermarle Sound and the various *Weapemeoc* Indian villages. The rest of the party briefly examined the Chesapeake and returned to England.

The principal river seen in White's Albermarle Sound is the Roanoke (not named). Located on its northern banks is the village of *Moratuc,* home of the Morotoc (Algonquian) Indians. The river originates further west in mountains of a region labelled *Mongoack* for the Mangoak Indians who, it was rumored, possessed a curious metal known as *wassador* which was said to be the color of copper but paler and softer. Other accounts placed this land of strange metal as lying yet further inland. The Europeans did not escape the temptation to believe that these were veiled reports of gold.

Though the lure of gold had remained seductive in the European psyche, the Roanoke settlers were lax in the quest for self-sufficiency. Falling into an oft-repeated pattern, Indian generosity and cooperation were relied upon for food supplies. Indian stores and Indian patience, however, were inadequate for both peoples and food supplies were soon depleted. An expedition which went up the Roanoke in the Spring of 1586 in search of the Mangoak's enigmatic yellow metal found the Indian villages deserted. Their inhabitants had fled inland with their staples, and only the chief of Roanoke Island itself continued to help support the British settlers.

Some of the colonists were sent thirty miles south to Croatan Island *(Croatoan),* which forms part of the Outer Banks itself and thus affords an unobstructed view of ocean. On June 8 of 1586 word was received from the Croatan party that twenty-some sails had been sighted. Sir Francis Drake, en route back to England from his brutal West Indian raids, had come to offer supplies or passage to the colonists. All opted to return to England. Shortly afterwards, Grenville returned with supplies and, finding the place abandoned, left fifteen men to guard Roanoke's fort.

In July of the following year a fresh group of colonists set out to fetch the Roanoke guards and then sail north to begin a new colony in a more favorable location along the Chesapeake Bay. John White was their governor. Grenville's garrison had not survived, however, and the relocation did not transpire. White helped familiarize the new colony with the region, and introduced them to the still-friendly Indians of Croatan Island. He then returned to England to secure provisions, leaving behind both his daughter and newly born grand-daughter, Virginia Dare, the first child of English parents born in the New World.

But war with Spain delayed White's return to the colony, and even after England's defeat of the Armada in 1588, fear of Spanish reprisal dampened further overseas adventures for another two years. In 1588 Walter Raleigh also attempted to send aid, but the captain of the two ships which had been given the task was distracted by pirating ambitions and returned to England after a damaging bout with French ships.

It was not until 1590, when his map and paintings were engraved on copper and published by Theodore de Bry, that White was finally able to return to Roanoke. But the colonists had vanished, leaving only the letters "CRO" carved on a tree to indicate their fate. White followed the cryptic lead to Croatan Island, but bad weather prevented his landing. What became of the colonists remains uncertain.[274]

---

273. This is the origin of the modern name "Hatteras," but the geographic features have changed over the past four centuries.

274. Although the colonists are commonly assumed to have perished, there is also the optimistic view that they were peacefully assimilated into Indian culture. In the early eighteenth century, one John Lawson visited the Outer Banks and was told by Hatteras Indians of white ancestors.

An original manuscript of White's map such as that which de Bry used as his model is still extant.[275] This engraved version, however, both lacks and adds some nomenclature to that manuscript, containing three more place-names overall along coasts and rivers. These may have been added during the 1587 voyage.

As for Sir Walter Raleigh, the original champion of Chesapeake settlement, by 1592 he had fallen into disfavor with Queen Elizabeth for his affair with (or secret marriage to) one of her maids-of-honor, then pursued a futile search for the riches of *El Dorado* in Guiana,[276] and in 1603 was accused of "treason." He lived long enough to see the Jamestown settlement take root, being executed a decade later in 1618. His execution was justified on vague and rather dubious charges, and may have been a matter of political convenience to win Spanish favor.

- - - - - - - - - - - - - -

275. In a manuscript album by White, presently in the British Library, consisting of two maps and 76 watercolor drawings.

276. About 1530 Spanish explorers were told of "El Dorado," a king gilded with gold who ruled a kingdom called Manoa on a Lake Parima.

40. MID-ATLANTIC COAST AND NEW ENGLAND

Norumbega et Virginia.

["Norumbega and Virginia"]. [in: *Descriptionis Ptolemaicae aumentum, sive Occidentalis notitia brevi commentario illustrata studio et opera Cornely Wytfliet Louaniesis.]* Cornelis Wytfliet, Louvain, 1597.

Medium : copperplate engraving.
Size of original : 235 x 300 mm.

PLATE : 26

Wytfliet's atlas, conceived as an addendum to Ptolemy, is the first specifically devoted to the New World. His map *Norumbega et Virginia,* focusing on the mid-Atlantic coast and New England, does not appear to be an exact derivitive of any one prototype, or at least of any surviving prototype.

Much of this map's inspiration comes from John White's map of Virginia (entry 39). Wytfliet borrowed that work's Carolina and Virginia nomenclature, the chain of islands forming the Outer Banks, its primitive Chesapeake Bay, and large Albemarle Sound. He however tempers the angularity and orientation of White's geography with the softer curves of Mercator, and relies on Mercator for the region of *Norembega.* In adapting the new White data to the Mercator shell, Wytfliet misplaces the coast to the north. Albemarle Sound, which appears at about 38° latitude on the 1585 White manuscript, already two degrees north of its actual location, is now shoved up to 40°, in line with what is now Philadelphia. The Chesapeake Bay, placed at above 40° by White, is pushed by Wytfliet to nearly 43°, the latitude of what is now southern Maine.

The mythical realm of Norumbega lies to the north, with its reputedly fabulous capital city of the same name lying along the banks of a "great river" *(R. grande).* Soon, in 1604-07, Champlain would search for Norumbega as one of the goals of his exploration of eastern Canada. In the wake of his failure to locate the kingdom, Norumbega would soon face extinction. It virtually disappeared on maps engraved after the first quarter of the seventeenth century.

Past *Norembega,* the coast extends due east (rather than northeast), thereby compensating for the erroneous latitudes of the White data to such an extent that by Cape Breton Island the error has reversed, that island being plotted slightly too far south. Both these problems —— the tendency to chart the northeast coast in an east-west orientation, and the plotting of many landfalls too far north —— were common flaws in maps of the sixteenth century. Both were the result of magnetic declination of the mariner's compass at high latitudes, a phenomenon which was particularly pronounced in American waters at that time. In eastern Canada, Wytfliet again deviates from the Mercator prototype, where the large gulf of a *R. Primero* ("First River") is seen, and Cape Breton Island lies quite alone far out in the Atlantic.

The fact that Ptolemy's *Geographia* was used as the vehicle for this atlas, though based on knowledge post-dating Ptolemy by one and a half millennia and concerning a region of the earth entirely unknown to him, illustrates the reverence with which classical knowledge was held throughout the Renaissance.

- - - - - - - - - - - - - -

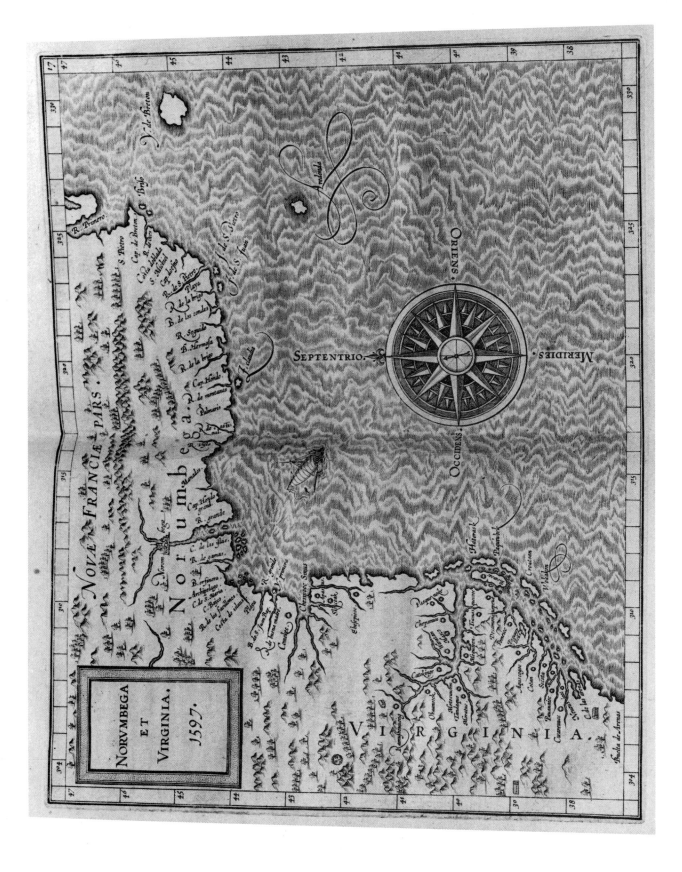

PLATE 26

Cornelis Wytfliet
Louvain, 1597

(entry 40)

41. NORTH AMERICA

America Pars Borealis, Florida, Baccalaos, Canada, Corterealis.

["The northern part of America, {being} Florida, Baccalaos, Canada, Corte-Real"]. Cornelis de Jode, Antwerp, 1593. Map of North America from the *Speculum Orbis Terrae.*

Medium : copperplate engraving, with original hand color.
Size of original : 370 x 505 mm.

COLOR PLATE : XIX

The French, like the British, resented the exclusive Iberian franchise on new lands dictated by the papal demarcation line, and tried to stake out claims of their own in what is now the United States. They targeted the region of northern Florida, officially in the Spanish sphere of influence. Although it had been abandoned by Spain as a region of serious exploration by a decree of Philip II in 1561, French presence there was nonetheless viewed as trespassing.

The experience of France's first attempt to colonize northern Florida, in 1562, is sadly parallel to what would happen to the British in Roanoke two decades later. Jean Ribault and René Goulaine de Laudonnière commanded the mission, which crossed the Atlantic directly from the French port of Havre de Grace to North America, rather than sailing the more common route through the Spanish Main and north. De Jode has written *Laudnner huc. appulit* and *Ribaldus huc.* at their landfall on the Florida peninsula. From there, they reconnoitered north, reaching a large river they named *Mai* because they found it on the first day of May. This is the St. Johns River, visible on de Jode's map as *R Mayo* near a region called *Tierra de las Pinas*. Still further north they encountered the Sea Islands which hug the coast of what is now South Carolina, and were impressed by the safety those islands afforded them. The largest sound along that coast they dubbed Port Royal, which name it has retained until today. De Jode latinized it as *P. Regalis*. On the island now known as Parris Island in Port Royal Sound, a fort and colony was set up which they called *Charlesfort,* spelled *Charlefort* by de Jode and indicated slightly north of *P. Regalis*. Ribault returned to France, promising to secure supplies for the little garrison of thirty settlers. But, as with John White in Roanoke, war back home prevented his timely return, in this case the civil war between the Protestants and Catholics. Ribaut fought for the Protestant cause in Dieppe, fleeing to England when Catholicism triumphed. While in England, he flirted with switching allegiances and claiming his prior exploits in the name of Queen Elizabeth. In the interim, the pitiful survivors of his American colony had constructed a boat, attempted the crossing back home, and were rescued by an English vessel.

In 1564, a new French expedition, commanded by Laudonière, established a settlement called Fort Caroline near the River *Mayo* discovered on the earlier expedition. Among their crew was Jacques Le Moyne, who was both to chart and illustrate the region. Although the colony fared well initially, they mistreated their neighboring Indians and yet relied upon them to supply their food, a combination which rendered the colony's existence precarious. A successful bid by the Spanish to oust them soon put a final and bloody end to Charlesfort. Laudonnière and Le Moyne were among the few to escape. Having safely returned to France, however, by 1582 Le Moyne had fled to England to escape the Huguenot presecution. It was in England that Theodore de Bry became aware of his marvelous drawings, map, and narrative. He tried unsuccessfully to purchase them from Le Moyne in 1587, but the following year Le Moyne died, and de Bry then bought the material from his widow. He published it in 1591 as the second part of his mammoth project of compiling accounts of voyages, following the publication of John White's work the year before.[277] Also in 1591, a map-seller of modest success by the name of Gerard de Jode died.

Gerard had been in competition with Abraham Ortelius; an atlas de Jode had introduced in 1578 flopped in the shadow of Ortelius' popular *Theatrum,* introduced eight years earlier.[278] Upon Gerard's death, the firm was left in the hands of his widow, and his son Cornelis. In 1593 they again tried to win a share of the atlas market, introducing a new atlas which contained both some of the father's maps as well as new maps. This map of North America is an example of the latter. Although this effort fared somewhat better than the 1578 work, it still sold too poorly to warrant another edition.

De Jode used de Bry's engraved versions of John White's and Jacques Le Moyne's maps to chart the southeastern part of his map of North America. White's Chesapeake is placed between 41° and 42° north latitude (present-day southern New England), and the nomenclature of John White, Verrazano, and Estavão Gomes intermingle. *C. de las Arenas,* the prominent cape on the mid-Atlantic coast, is often construed as being Cape Cod, but cannot be so if early Spanish prototypes, which clearly place it along the part of the coast scouted by Ayllón, are accurate. The place-name de Jode has matched it with, however, is from the voyage of Gomes, who did in fact pass Cape Cod.[279] This instance exemplifies the problem of identifying land features when nomenclature and geography may be mis-matched. North of it, reflecting England's Roanoke adventure, lie the Indian villages of *Secotan* and *Pomerock,* and the town and bay of *Chesapooc* (Chesapeake). *Norombega,* our New England, is shoved far over nearly due east. The city of Norombega lies on the banks of Gomes' "River of Deer" *(Gamas).* De Jode inscribed several historical annotations on the map, but mistakenly ascribes the date of 1529 for Verrazano's principal voyage (actually 1524) and 1507 for Cabot (actually 1497).

In addition to White's geography, de Jode used his illustrations of Virginia Indians to supplement the otherwise empty space in the Atlantic. From left to right they are a warrior prince, a chief lady of *Secota,* a holy man of *Secota,* a *Dasemonquepeuc* woman carrying her child, an aged man of *Pomeiooc* in winter clothes, and a young noble woman of *Secota.*

Unobstructed ocean north of the continent alludes to a passage around America to the Pacific. Inland from the northern coast a large Lake *Conibas,* said to be fresh water, contains an island with the city of *Conibaz.* It is not known whether this lake, which is found as early as the mid-century as *Coniuas lacus,* represents early knowledge of Hudson Bay (or the Great Lakes) based on indigenous sources, or is entirely fanciful.[280] It is similar to the *Mare Dulce* ("fresh sea") appearing on such maps as those of Mercator.[281] To the east of Conibas is Cartier's purported kingdom of *Saguenai.*

The map's peculiar top-heavy look exaggerates the higher latitudes: enigmatic Lake Conibas, the Arctic, and Cartier's discoveries in Canada dominate. This is a result of the distortion of the map's projection, which stretches the geography onto a grid of parallel

---

277. The story of the expedition was first published by Ribaut in London in 1563. Le Moyne's original pictures have been lost except for one which was discovered in 1901 (now in the New York Public Library).

278. Koeman (Atlantes Neerlandici, Vol. II p. 205) quotes Denucé in speculating that Ortelius "pulled strings" to prevent de Jode's receiving the royal imprimatur necessary to publish the atlas, this being the reason why Gerard, eighteen years Ortelius' elder, was so late in the atlas project.

279. According to Sauer, Gomes' "cabo de las arenas" is Cape May. See also page 86 in entry 24 (Gastaldi).

280. See Harrissee, The Discovery of North America, map 253 (p. 645-646). "Conibas" might possibly be a corruption of "Caniba" (i.e., Cariba, Caribales).

281. An incidental point supporting Conibas' being real rather than mythical is the fact that a separate map of the region published by Wytfliet in 1597 is known in two states, with altered nomenclature, suggesting that Wytfliet was juggling actual data. The Arctic map of Mercator (1595) has both Conibas and Mare Dulce.

meridians, inspired by Gerhard Mercator's 1569 map. Unlike a true Mercator projection, however, de Jode does not progressively increase the latitudinal distances to maintain equality between a true and plotted straight line. As with Mercator, the entire upper border of the map represents the single geometric point of the North Pole. The depiction of the Arctic as four major islands (of which parts of two appear) was also borrowed from Mercator.

In de Jode's Atlantic, vestiges of semi-mythical lands of medieval times still cling to life. Two are reminiscent of fifteenth century portolan chart data, possibly inspired by early Portuguese Atlantic voyages, and were not commonly used by mapmakers in the sixteenth century. One of these is *Santana*, the "Satan" island typically found as a large northerly island on some charts beginning in Prince Henry's time; the other is *Sept Citez*, the island of the Seven Cities which became nearly synonymous with the island of Antilia by the end of Henry's time.[282] De Jode places them in roughly the same relative position to America they were commonly given to Asia a century and a half earlier, *Sept Citez* lying due east of Florida (in between the two groups of White's Indians) and *Santana* lying to the north.

The waters of de Jode's northern Atlantic are largely derived from the purported adventures of the Italians Nicoló and Antonio Zeno in 1380. The story and map of their doubtful fourteenth century voyage was published in Venice by a descendant in 1558. As Purchas relates it, the brother Nicoló, *"being wealthy, and of a haughtie spirit, desiring to see the fashions of the world, built and furnished a Ship at his owne charges,"* passed through the Strait of Gibraltar and was promptly carried astray by a tempest.[283] They were adrift until reaching de Jode's *Frislant*, where they were saved from barbarians by a Prince Zichmui who *"spake to them in Latine, and placed them in his Navie, wherewith hee wonne divers Ilands."*

Following *"divers notable exploits"* they reached *Groenlant* (*Engronelant* on the Zeno map), *"where hee found a Monasterie of Friers of the Preachers Order, and a Church dedicated to Saint Thomas."* This monastery was situated by an active volcano, which afforded a hot spring whose water was used for heating and cooking. The brother Antonio then reached *Estotilant*, assimilated by de Jode as part of easternmost Canada. Next he sailed south to *Drogco* (Zeno's *Drogio*), during which voyage cannibals were encountered. Among the other islands they reached was de Jode's *Icario* (west of *Thule*), where there were *"Knights thereof called Icari, descended of the ancient pedigree of Dedalus, King of Scots."* De Jode substitutes the archiac *Thule* for Zeno's *Islanda*.

Frislant, which appeared on charts as early as 1500, was possibly based on early knowledge of Iceland, and certainly serves as Iceland on de Jode's map.[284] Both Greenland and Iceland, then, appear in duplicate: *Groclant* and *Groenlant* are both Greenland, and *Thule* (the Shetland Islands of Ptolemy) and *Frislant* are Iceland. From other northern traditions, de Jode has included the legendary Irish island of *S. Brandain*.

- - - - - - - - - - - - - -

---

282. In entry 9 (Beccari), see page 33 for "Satan" island and Antilia, and page 43 for Seven Cities.

283. Samuel Purchas, Purchas His Pilgrimes, part III, p. 610-11.

284. See Tony Campbell, "Portolans Charts from the Late Thirteenth Century to 1500," in Harley/Woodward, The History of Cartography, p. 414.

42. NEW ENGLAND

New England, the most remarqueble parts thus named.

John Smith, London, 1616 (circa 1627). State VI of nine).
The map first appeared in Smith's *The Description of New England,* 1616.

Medium : copperplate engraving.
Size of original : 300 x 355 mm.

PLATE : 27

Captain John Smith was an adventurous and fearless sort who led a colorful, if now largely romanticizied, life. Though brazen and hardly modest, Smith's accomplishments were significant, and his undaunting perseverance contributed greatly to the success of the British empire's grip on North America.

Following the death of his father in 1596, Smith abandoned his employment as a merchant's apprentice to brave the seas. Enduring a novel's share of misadventure, he fought in the wars against the Turks in Hungary and Transylvania, escaped enslavement in Turkey, and eventually returned to England. From there in December of 1606 he set sail for the New World with London's Virginia Company. The following April, reaching a site which lay north of both Roanoke and of where the Spanish colony of San Miguel had failed nearly a century earlier, they struck camp and thus founded their settlement.[285] They named it Jamestown in honor of their sovereign, James I.

Smith, immensely resourceful, clever, and persuasive, was instrumental in the ultimate survival of Jamestown and thereby ensured England's foot-hold in the New World. He helped keep order in volatile times, sought sources of food, and surveyed the Chesapeake. While exploring the bay Smith got into mortal trouble with the Indian chief Powhatan, from which he was purportedly rescued by the king's daughter in one of history's great romantic episodes. This "princess," Matoaka *(Pocahantas),*[286] herself became another example of Smith's influence, as it was through the Smith incident that her influencial involvement with European civilization began, becoming a diplomatic link between the two cultures, and marrying John Rolfe, whose career heralded the tobacco trade with England.[287]

Smith returned to England in 1609 with wounds received when a gunpowder pouch he was carrying exploded. But five years later a group of London merchants headed by Ferdinando Gorges hired him to sail once again to the New World, this time to explore the "north part of Virginia." During six weeks of June-July 1614, Smith charted the New England seaboard and dabbled in trade. He returned with a wealth of fish and furs, with surveys for this map, and with a mad passion for the British settlement of the region.

---

285. But see footnote 222.

286. The English perceived her as a princess because her father was a king, but in fact their society was matrilinear. "Pocahantas" was a nickname. It is most commonly interpreted as "little wanton" (for her personality), sometimes as "the outcast" (for her sympathy for the Europeans), or it may instead be a form of the Algonquian word for the male genitals, accorded her as a nickname for her flirtatiousness (according to Sale, using Strachey's 1612 Algonquian dictionary for the word "pocahaac"). Her saving of Smith's life may possibly have been a premeditated ploy by her father for him to maintain the appearance of being a strong leader with his people while avoiding retaliation from the British. Her marriage to Rolfe was probably a political convenience to establish some stability to the two peoples' warring relations, thus importing to the New World a device which had long been politics-as-usual in Europe. Later in the seventeenth century the British established laws to ban Anglo-Indian marriage.

287. Ironically Rolfe's sovereign, James I, fought the introduction of tobacco as a health hazard. Even before the Jamestown settlement he denounced tobacco in a pamphlet entitled A Counterblast to Tobacco (1604).

His map was not printed upon his return from this mission. Rather, he occupied himself with two unsuccessful attempts to return to New England, on the second of these being taken prisoner by a French pirate ship. But, the fabulous and nearly surrealistic quality of his life never waning, Smith maneuvered a daring escape and returned to England. The year was now 1616. It is not known where his manuscript charts of the New England coast were during these two years since his return, nor even whether he at this point had them at his disposal, but a map was prepared and engraved on copper. It was intended to promote settlement in New England.[288]

That the Pilgrims used a copy of Smith's map during their fateful crossing of 1620 is clear evidence that it was successful. The map was the best available of the region between Cape Cod and Penobscot Bay, and became the foundation work of New England cartography. Originally, the map contained no English names; Smith, however, asked the then fifteen-year-old Prince Charles to change "barbarous" Indian names to Anglo and Scot names *"for such English, as Posterity may say, Prince Charles was their Godfather."* [289]

The Pilgrims adopted the map's name *Plimouth* for their colony; only this name, *Cape Anna* and *The River Charles* still survive in their locations shown by Smith.

Cape Cod is dubbed Cape James in honor of James I. The term "Cape Cod" had, however, already been given to the claw-shaped protrusion by Bartholomew Gosnold for the enormous quantity of cod in its waters noted during his voyage to "the north part of Virginia" in 1602. While Prince Charles favored Anglo-Scot names for posterity, posterity ultimately favored the recurrent *Baccalao* (codfish) theme over English kings.

- - - - - - - - - - - - - -

288. Some authorities (e.g., Brown) question Smith's claim that he himself made the surveys upon which the map of New England was based; the same doubts have been raised about the Virginia map. Smith acknowledged having consulted with the Indians to supplement his data when composing the Virginia map ("the rest [of the map beyond cross marks] was had by information of the Savages, and are set downe according to their instructions"), although the New England map does not extend as far into the interior where tapping Indian data would have been most useful. For the present map, Smith himself, in <u>A Description of New England</u> (p. 4-5), seems to acknowledge his reliance on existing charts, stating that he "had six or severall plots of those Northern parts, so unlike each to other, and most so differing from any true proportion, or resemblance of the Country, as they did mee no more good, then so much waste paper, though they cost me more."

289. From the dedication in <u>A Description of New England</u>.

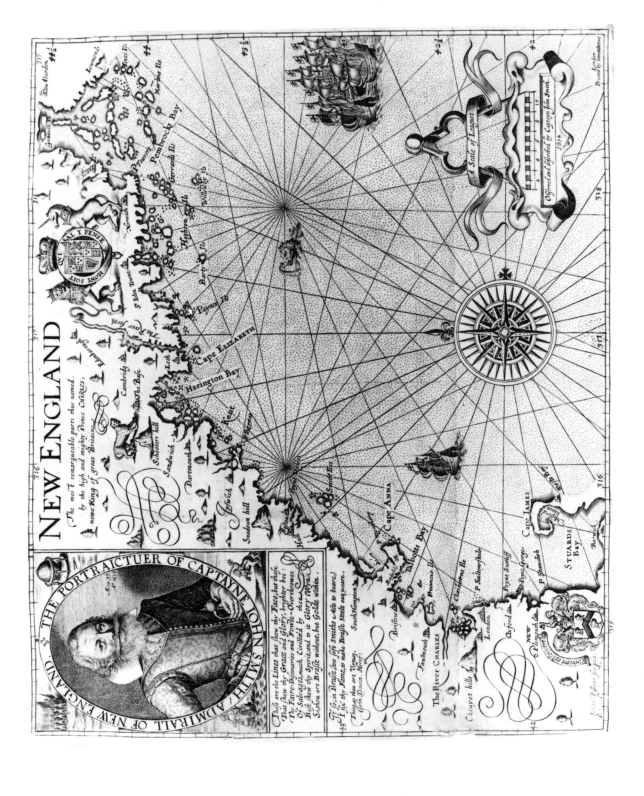

PLATE 27

John Smith
London, 1616
(circa 1627)

(entry 42)

43. PORTOLAN CHART

[untitled portolan chart of the Mediterranean and Black Seas].

Nicholas Comberford, London, 1626.

Medium : illuminated manuscript in vellum.
Size of original : 405 x 765 mm.

COLOR PLATE : XX

Along with England's freshly invigorated interest in overseas adventure, the tide of which John Smith brilliantly harnessed, came higher stakes in a domestic chart industry. Although the precedent and success of Waghenaer's printed sea-atlas in Holland in 1583 closely coincided with England's new exploits, the British did not develop a strong trade in printed charts until the third-quarter of the seventeenth century. Local demand for charts was instead filled by a loosely-knit school of chart-makers which had developed in London towards the end of the sixteenth century. It has become known both by the term *"Thames School,"* because of its makers' invariable location along that river, and *"Drapers' School,"* owing to the apparent association with the drapers trade of many of its members, including the maker of this chart, Nicholas Comberford.[290]

The present chart is Comberford's earliest known surviving work.[291] It is on vellum, and had formerly been pasted on hinged wooden boards to facilitate its use and storage both aboard ship and at dockside facilities; the holes visible along its centerfold were cut to allow the hinges to protrude. Such charts were called *platts*.

This "platt" covers the area from the Strait of Gibraltar through the eastern bounds of the Mediterranean, and includes most of the Black Sea. It is representative of the portolan charts which a British sailor —— such as John Smith —— might have purchased from a "platt-shop" when embarking on a voyage to the Mediterranean or Levant during Britain's heyday of exploratory zeal. As the chart's function was strictly maritime, the use of inland data was unnecessary. Instead, lists of islands occupy Europe, and a mileage scale is placed in Asia Minor. Atypically, it is not criss-crossed with the rhumb-lines which formed a framework on most sea charts. In Africa lies Comberford's inscription, which reads *"This platt was made by Nicholas Commerford . . . for John Gibbons. Ano 1626."*

Although little is known about Comberford himself, record exists of a stranger's visit to his shop in 1655. From the account it is evident that the old man, his family and daughter-in-law were living in deplorable conditions. That this was sadly typical of the chartmaker's lot is suggested by the fact that the visitor, who had sought out Comberford believing that he was a relative from whom an inheritance might be in order, *"began to fear for the success of* [his] *journey"* upon hearing from townspeople that the man whom he sought was a chart-maker.[292]

As was traditional, Comberford passed his craft on to the next generation. Among his pupils was one John Burston, who in turn mentored a talented apprentice named John Thornton. Thornton would become hydrographer to both the East India and Hudson Bay Companies, and help begin the printing of sea charts in England, thus speeding, ironically, the demise of his country's manuscript portolan chart trade (see entry 55).

---

290. See Tony Campbell's "The Drapers' Company and its school of seventeenth century chart-makers" in <u>My Head is a Map</u>, and Thomas Smith's "Manuscript and Printed Sea Charts" in <u>The Compleat Plattmaker</u>.

291. This is the chart discussed in Thomas Smith's article "An Early Portolan of the Mediterranean by Nicolas Comberford, 1626" in <u>Imago Mundi</u> no. 29 (1977), and is also the 1626 Comberford chart he refers to in his "Manuscript and Printed Sea Charts" in <u>The Compleat Plattmaker</u>.

292. See Thomas Smith, "Manuscript and Printed Sea Charts" in <u>The Compleat Plattmaker</u>, p. 91-92.

44. NEW ENGLAND

[untitled map of New England].

William Alexander, London, 1624.
From *The Mapp and Description of New England.*

Medium : copperplate engraving.
Size of original : 250 x 350 mm.

PLATE : 28

The charter of April 10, 1606, which had formally established the Virginia Company, actually named two such companies: the Southern Colony of Virginia, and the Northern Colony of Virginia. The former, based at Jamestown and supported by wealthy adventurers, flourished ; the latter produced only the feeble and short-lived Popham Colony at Sabino.

On the third of November in the year 1620, under the reign of James I, a new charter was granted which superceded the defunct Northern Virginia Company. It was *The Council for New England,* established for the *"planting, ruling, ordering, and governing of New England, in America."*[293]

In contrast to the earlier company, this enterprise was pursued with great vigor. A leading proponent was Sir Ferdinando Gorges, a man of privilege in royal circles and head of the businessmen who had sent Captain Smith to investigate the commercial prospects of New England. He had been intrigued with New England settlement since the Popham attempt.

Forty patentees were named and empowered to hold territory from north latitudes 40° through 48°, stretching westward to the Pacific Ocean. At the king's request a Scot, Sir William Alexander, had been one of the beneficiaries. In September of 1621 Alexander received a royal charter for all the land east of the St. Croix River on the Acadian Peninsula.

The political climate in England was different than it was when the Northern Virginia Company was put on paper. England had been without a parliament since 1614, freeing various people and associations with privileges from the Crown to pursue their own initiatives. Gorges exploited the freedom by securing for himself a monopoly on fishing in New England, resulting in long challenges by The Virginia Company.

Gorges had to contend not only with smoldering opposition to his fishing monopoly, but also with internal squabbling among patentees. In the summer of 1623, a meeting with the King was held in Greenwich at which the number of patentees was reduced to twenty. The following year Alexander published a book about New England which contained this map, locating the respective parcels of land awarded these twenty titleholders. They extend from Cape Cod (the *Cape James* of Smith's map has already been abandoned in favor of the earlier, modern name) through what is now northern Maine.

Alexander called his lot *New Scotlande* in honor of his country, or ultimately Nova Scotia, and he imported regional names from his homeland as well. He transposed the names *Twede* and *Clyde* from the Tweed and Clyde rivers in southern Scotland to denote the St. Croix and St. John Rivers (in present-day New Brunswick). Poetically, just as Scotland is partitioned from England by the firths of Solway and Forth, so does Alexander demark his New Scotland from its neighbors by rivers *Sulway* and *Forthe*, shown as a tributary of the St. Lawrence and an inlet in the St. Lawrence Gulf. The extension of mainland which is now Nova Scotia proper is dubbed *The Province of Caledonia* after the ancient Roman name for Scotland. He named southern Newfoundland after himself *(Alexandria),* although, unlike his namesake who tried to conquer the world two thousand years earlier, his name has not been enshrined by posterity.

---

293. Winsor, <u>Narrative and Critical History of America</u>.

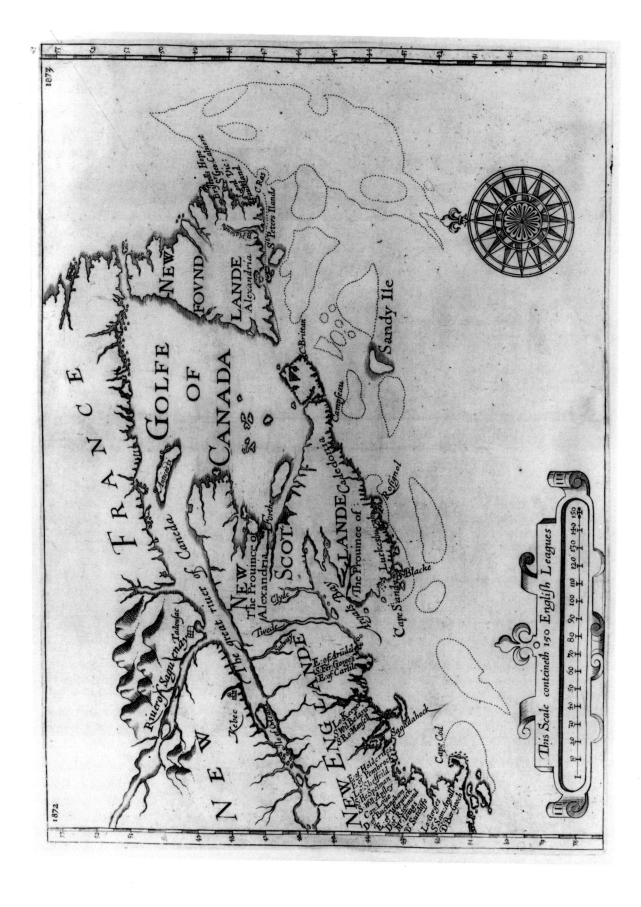

PLATE 28

William Alexander
London, 1624

(entry 44)

By the time Alexander's map was published in 1624, England had regained its parliament. In that year, with several British ships under arrest by the Admiralty for fishing off the New England coast in violation of Gorges' extravagant monopoly, Parliament passed a measure declaring the fishing restriction illegal, insuring that New England waters were to be open to all subjects of the Crown. The monopoly in which Gorges had vested his efforts was now dissolved. Alexander persevered in his dream of creating a Scotland in the New World, but the colonial enterprise ultimately failed.

- - - - - - - - - - - - - -

45. NEW ENGLAND

The south Part of New England, as it is Planted this year, 1639.

[in: *New Englands Prospect*].
William Wood, London, 1634 (1639).

Medium : woodcut.
Size of original : 305 x 185 mm.

PLATE : 29

Among the many British subjects lured to the New England colonies was William Wood, who settled in Lynn, Massachusetts, in 1629. Over the course of the next four years he recorded his observations of the numerous settlements which had been formally or unofficially established in the vicinity of Massachusetts Bay. During his stay, about a thousand colonists arrived as a result of the chartering of the Massachusetts Bay Company. The village of Boston was started by a group of these Puritans.

Wood returned to England in 1633 and published his book, which was the first topographical description of Massachusetts. Its accompanying small woodcut map, though crudely executed, was the finest depiction of the coast from Narragansett Bay through New Hampshire yet produced. It was also the first printed map of New England by an English settler, and the first to name Boston. Esthetically, it is novel for the peculiar way in which the coasts are shaded, producing the effect of relief.

As a map directly resulting from British presence, the Wood forms a link between the John Smith map of 1616, and the John Foster map of 1677. Wood's geographic information is believed to be related to a manuscript attributed to John Winthrop (circa 1633), governor of Massachusetts Bay, or possibly based directly on the Winthrop map in part (that map covers only the area of Massachusetts Bay and does not cover Cape Cod). In the vicinity of Boston, Wood records more than thirty Indian and English settlements which appear on a printed map here for the first time. A key on the Winthrop manuscript supplements the Wood data, for example commenting that at Boston is *"the Wyndmill, the fforte, and the Weere."*

Wood's book attempted, in the words of the titlepage, to "[lay down] *that which may both enrich the knowledge of the mind-travelling Reader, or benefit the future Voyager."* In the preface he states that he has described *"my dwelling place where I have lived these four years, and intend God willing to returne shortly again."* Although the author's fate is not definitely established, in 1635 a William Wood made the crossing from England to Massachusetts in the *Hopewell,* and he is generally assumed to have been the same man, fulfilling his promise made in the book's preface.

- - - - - - - - - - - - - -

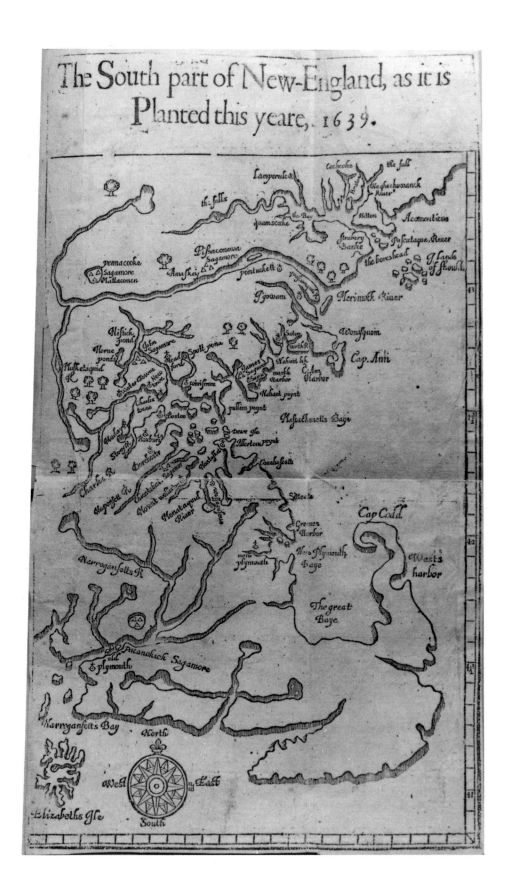

PLATE 29   William Wood, London, 1634 (1639)   (entry 45)

46. NEW ENGLAND

Nuova Belgica e Nuova Anglia.

["New Belgium and New England"]. Sir Robert Dudley, Florence, 1646 (1661). Chart of New England from Dudley's *Arcano del Mare.*

Medium : copperplate engraving.
Size of original : 300 x 400 mm.

PLATE : 30

One of the tricky aspects of ocean navigation was the fact that a plotted straight line on a sea-chart did not represent a straight line of sailing. In 1569 Gerhard Mercator introduced a map projection which attempted to correct this by intersecting all parallels and meridians at 90° angles and progressively lengthening latitudinal distance from the equator. The resulting distortion of this projection was, however, disconcerting to the uninitiated. Further, the precise calculation of distances on such a projection was tricky, and it was not until thirty years after Mercator's experimental map that an Englishman, Edward Wright, devised tables for this purpose. In the wake of his breakthrough, the chart makers of Dieppe, a French port on the English Channel, began adopting the Mercator projection about 1630. It was becoming increasingly clear that Mercator's brainchild presented a major advantage for navigators. The most influential resurrection of the projection came at the hands of Sir Robert Dudley, an Englishman in self-exile in Italy. His atlas, the first exclusively to use the projection, first appeared in 1646-47; the present chart comes from a second, corrected edition of 1661.

Dudley was an illegitimate son of the Earl of Leicester and brother-in-law of Thomas Cavendish.[294] A Catholic, Dudley had settled in Florence to escape religious persecution. He was well-regarded as a mathematician and navigator. The preface to his atlas states that the engraver Antonio Francesco Lucini spent twelve years in making the copper-plates, revealing that work on the project had begun about the same time the Dieppe makers had begun using the projection.

Dudley's work was significant for his decision to refine and use the Mercator projection, for his incorporation of winds and currents, and for its accuracy and detail. It is also revered for its calligraphy and its singularly clean, elegant design.

- - - - - - - - - - - - - -

294. Cavendish commanded the third circumnavigation of the earth (after Magellan and Drake).

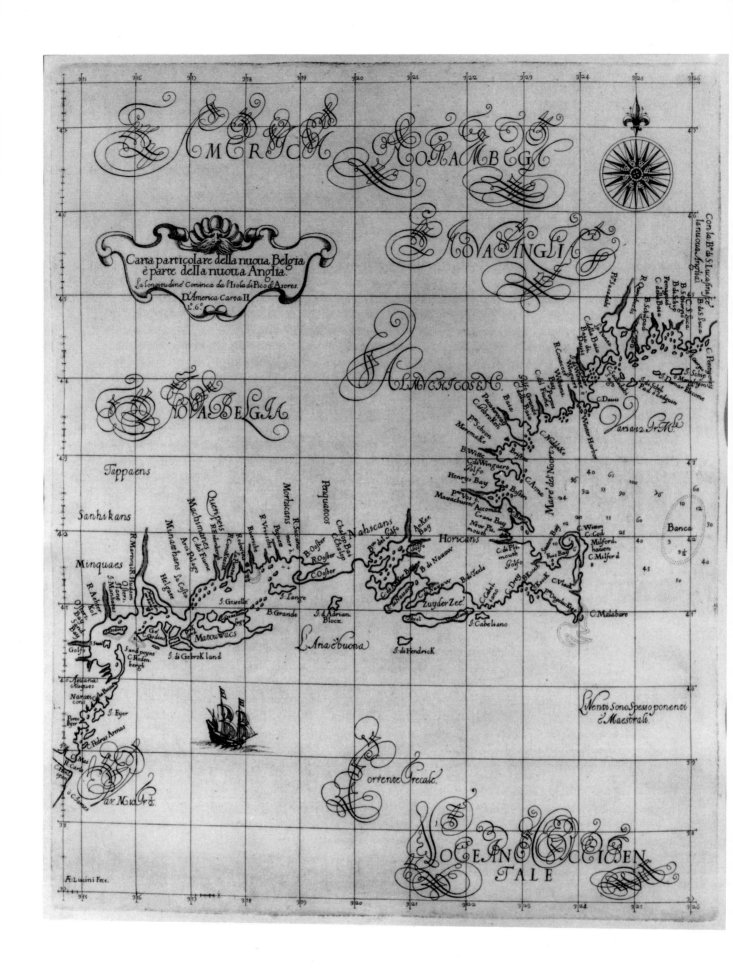

PLATE 30   Robert Dudley, Florence, 1646 (1661)   (entry 46)

47. NEW ENGLAND

A Map of New England, Being the first that ever was here cut, and done by the best Pattern that could be had, which being in some places defective, it made the other less exact; yet does it sufficiently shew the Scituation of the Country, and conveniently well the distance of places.

Map of New England from William Hubbard's *A Narrative of the Troubles with the Indians in New-England* . . . John Foster, London (and Boston), 1677.

Medium : woodcut.
Size of original : 295 x 385 mm.

PLATE : 31

In 1629 a charter was granted to the Massachusetts Bay Colony which set her northern boundary three miles north of the Merrimack River. But that river had been only superficially explored, and so the colonists, certainly aware of the overwhelming tendency of the earth's rivers to flows north-south and of the beneficial effect such a course could have on the extent of their land claim, sent several expeditions to follow its course. The Merrimack was in fact found to veer sharply to the north, and in 1652 its source was located at Lake Winnipesaukee. The colony then claimed a northern boundary running three miles north of the river's initial flow from the lake, dramatically expanding their domain and infringing upon the claims of Maine.

This rapid expansion of British settlement strained relations not only among the various colonies, but also between the European settlers and their indigenous hosts, the Indians. The principal Indian nations in New England, the Narragansetts and Wampanoags, were long tolerant, and even receptive, to British presence, both because they benefited from the European appetite for furs, and because it shielded them from the unfriendly Pequot Indians. But eventually both Indian nations felt that the British encroachment had become intolerable, and so in 1675 they rose against the colonies. The bloody war that ensued became known by the unlikely name of the Wampanoag's chief, King Philip.

Foster's map accompanied an early account of King Philip's war.[295] It was based on a survey done by William Reed in 1665, and served as a guide to the various battles and sites of British-Indian conflict. The map also lent legitimacy to the recently inflated boundaries claimed by Massachusetts: two parallel lines establish the colony's turf, the northern one plotted through Lake Winnipesaukee, gobbling up much of Maine, and the southern boundary conflicting with the Plymouth Colony's 1629 charter. A compromise partitioning of the Plymouth boundary, reached in 1664, is also marked however, shown as a diagonal line starting near Scituate *(Scituat)*.

In many respects (such as the charting of the Connecticut and Merrimack Rivers) Foster's map is an improvement over contemporary Dutch maps. Over sixty European settlements are indicated, most being identified by numbers.

Two nearly identical woodblocks were made for this map. The first, printed in Boston by John Foster, is regarded as the first map printed in America. The second was re-cut in London later the same year from a proof of the Boston block and was printed by Thomas Parkhurst. Several variations differentiate the Boston and London issues, but they are

---

295. The author of the book <u>A Narrative of the Troubles with the Indians in New England</u>, William Hubbard, was born in Essex County, England, in 1621. As a young man he accompanied his father to the New World, arriving in New England at the age of fourteen. He graduated from Harvard University in 1642; forty-four years later he was appointed acting President of Harvard. For most of his life he resided in Ipswich, serving there as minister of the village's church.

commonly referred to by their labelling of the White Mountains, shown correctly as *White Hills* on the Boston printing but as *Wine Hills* on this, the London issue.[296]

The most compelling evidence to show that the *Wine Hills* is in fact the copy is found in the different media with which the two produced words. The engraver of the Boston woodblock used inserted moveable typeset to produce many of the place-names, a practice which sometimes resulted in sloppy register of the letters. The London block, although its letters were cut entirely out of the wood, imitates the alignment of the Boston block's typeset. Probably a proof of the Boston block was shipped to England, put face-down on a fresh woodblock, and then varnished to render the paper translucent so the image could be copied. The reverse image seen by the woodcutter was not sharp, however, and the quality of the copy was further limited by any imperfections in the original pull. From these factors errors were introduced into this London issue. For example, Martha's Vineyard, clearly shown as *Martins Vineyard* on the Boston work, confused the London engraver, who took the "i" of *Vineyard* to be the right vertical stem of the letter "N," resulting in *Nnyard*. The Boston map's *Newport* is *Newper* on this London rendition, *Ipswich* is *Ipswieh, Casco* is *Caseo, Hamton* is *Haniton, Lancaster* is *Lancasten,* and *Sey-brook* is *Sey-byook*. The town of *Chensford* (Chelmsford) on the Boston issue became *Chensforo* on this London strike because the vertical line of the "d" was mistaken by the woodcutter to be the easterly of two lines representing the adjacent river.

The pioneer printer of the American printed map, John Foster, was described as an "ingenious mathemetician and printer" on his epitaph. He had begun printing at "the Sign of the Dove" in Boston in 1674, the very year in which a decree by the General Court which had forbidden any printing in Massachusetts Bay except Cambridge, dating from 1664, was repealed.

- - - - - - - - - - - - - -

---

296. Once the focus of dispute, David Woodward convincingly sorted out the relative dating and place of issue of the two. See "The Foster woodcut map controversy: a further examination of the evidence," in <u>Imago Mundi</u> XXI (1967), p. 52-61.

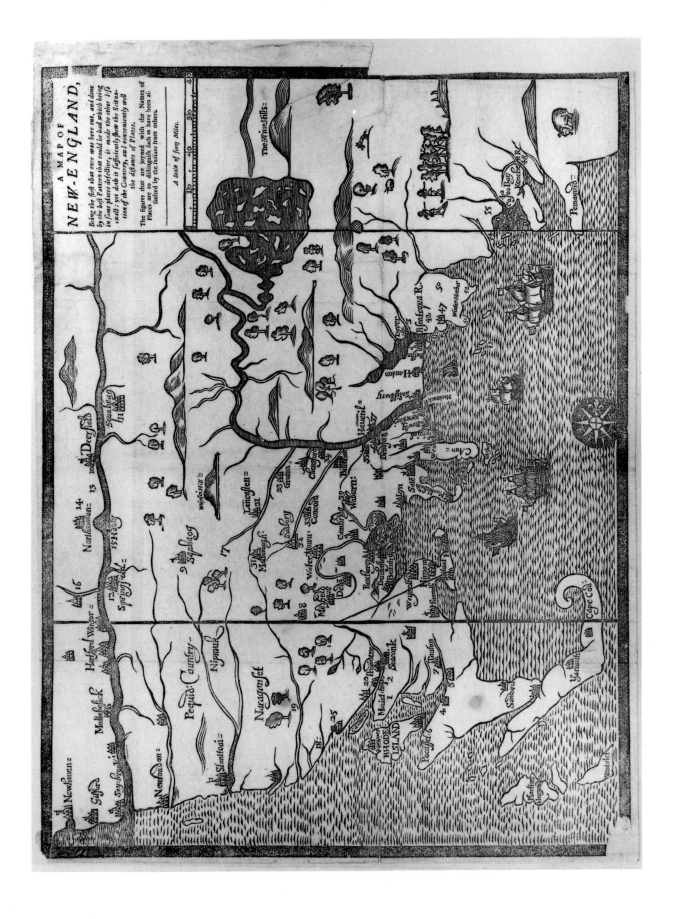

PLATE 31

John Foster
London, 1677

(entry 47)

48.   AMERICA [Wall Map]

L'Amerique Divisee Selon l'etendue de ses Principales Parties et dont les Points Principaux sont Placez sur les Observations de Messieurs de l'Academie Royale des Sciences. Dressee Par N De Fer, Geographe de monseigneur le Dauphin. A Paris, Chez l'Autheur dans l'Isle du Palais sur le Quay de l'Horloge a la Sphere Royale 1698.

["America divided according to the extent of its principal parts, and in which the principal locations are placed according to the observations of the Royal Academy of Sciences"]. Nicolas de Fer, Paris, 1698. Separate publication. Text panels added circa 1703.

Medium : copperplate engraving.
Size of original : 1,065 x 1,170 mm.

PLATE : 32

By the early years of the seventeenth century, the British and French had staked claims on American soil which were to determine colonial struggles for the next two centuries. Their footholds had been won, however, with an incomplete knowledge of the land's virtues, and so the sobering rite of exploring their prizes followed. For the British this was a sometimes rude morning-after, for although they had secured the mid-Atlantic coast with its milder climate and vast seaboard, it was constrained by the inland mountain ranges which run roughly parrallel to it. The French, conversely, who had won only the rugged coasts of eastern Canada, found themselves in control of extensive inland waterways. By the end of the seventeenth century they dominated the St. Lawrence and Great Lakes, which traversed nearly half the continent east-west, and the Mississippi River, which clinched the continent to the warm coast of the Gulf of Mexico, a back door to Spanish strongholds in Mexico and the Caribbean. As a result the French controlled the most strategic transportation and communications channels in America, and thus ruled the heart of the continent. The British found that they may have grabbed a prettier but less consequential piece of the pie, and this discontent set the arena for the violent French-English warring of the eighteenth century.

As France's grasp on North America solidified in the seventeenth century, so did her map-making mature. Important maps had been published in France during the fifteenth and sixteenth centuries, but her entry as a serious contender in the commercial map trade did not begin until the introduction of the atlas of Nicolas Sanson in 1650. Although the maps produced in France often lacked the decorative appeal and finesse of their Dutch counterparts, they excelled in recording the latest discoveries of her explorers. And as France had superior access to the least understood parts of North America, her maps became the most valuable and influential source of such data through the end of the colonial era.

France's rapid advances began in the early years of the seventeenth century when Samuel de Champlain investigated the St. Lawrence which Cartier had pursued in the mid-1530's. Champlain and many of the Jesuits and trappers who followed him believed that this vast waterway might offer a passage through the continent to the Pacific Ocean, thus giving fresh impetus to the unrequitted search for an easy sea route to China. This helped establish the fur trade over a large area, opening the way to the Great Lakes by way of the St. Lawrence and to the region north of the St. Lawrence by way of the Saguenay and St. Maurice Rivers.

Indigenous reports of waters lying to the west also nutured further exploration. The Huron Indians told Jean Nicollet, colleague of Champlain before the latter's death in 1635, of a vast sea to the west. Nicollet hoped this would be the Pacific; the quest brought him to Lake Michigan. Similarly vague reports told of a great river not far to the west, possibly the Wisconsin or Mississippi, but again construed as a water route to the western bounds of the continent. In 1673 Louis Jolliet and Jacques Marquette, following such leads, reached the Mississippi and Missouri Rivers (de Fer's *Oubache Riviere*). France's web was completed when

PLATE 32   Nicolas de Fer, Paris, 1698   (entry 48)

René-Robert La Salle explored the Mississippi fully to its mouth in the Gulf of Mexico, neatly establishing French control over the bulk of North America. French commercial interests and colonial designs now spread to the five Great Lakes up through Hudson and James Bays, extending by the Mississippi to the vast, temperate lands which lay directly west of England's settlements. The British were boxed in to their constricted seaboard claims.

This large, separately-issued map of Nicolas de Fer represents the culmination of French progress at the close of the seventeenth century in her bid to transform North America into a "New France." Improving on his immediate predecessors, he drew heavily on manuscript maps of J. B. Franquelin who, stationed in Montreal, compiled state-of-the-art data from various explorers. De Fer records the Great Lakes and Mississippi in better detail than Coronelli (1690), and eliminates Coronelli's erroneous placement of Taos south of Sante Fe (albeit by not showing it at all). He shows a better knowledge of the various Mississippi tributaries and of the lakes in central Canada west of Hudson Bay, and demonstrates a more complete understanding of the water systems connecting the St. Lawrence with Hudson Bay. North America appears as if it were a leaf, with the vast river systems its veins. What is so remarkable about de Fer's map is that it records how within a mere few decades France's explorers, traders, and missionaries had dragged Europe's knowledge of a vast stretch of the North American interior —— about a third of the continent —— from that of ignorance to sophistication.

Not all developments were positive. De Fer resisted the influence of Louis Hennepin, a Friar who discovered the Falls of St. Anthony (de Fer's *Saut St. Antoine*) and who published two maps preceding de Fer's (1683 and 1697). This was lucky for the Great Lakes, as Hennepin's delineation was regressive, but unfortunate for the Mississippi, which Hennepin had located quite accurately on his map of 1683 (although he did not show its southern section or mouth at all). Instead de Fer, following La Salle, places the Mississippi far to the west so that its delta lies in what is now Texas (an arrangement also adopted by Hennepin in 1697). The inability to determine longitude accurately was of course partly to blame for the misplacement, but it is probably more true that the inability to determine longitude accurately allowed La Salle to *deliberately* offset the river. By placing the Mississippi closer to Spanish claims in the Southwest, La Salle both fattened the breadth of continent claimed in the name of France, and made its banks appear to be an opportune base from which to pillage Spanish gold, thus encouraging French fortune seekers to flock to it. Not until the 1703 map of Guillaume de l'Isle was the Mississippi's longitude accurately set. That map of de l'Isle, however, would incorporate a bogus "long river" connecting the Mississippi with a salt lake to the west, introduced by the Baron Louis Lahontan in 1703, a temptation which de Fer in this earlier map was spared.

De Fer capitalized on the psychological value of maps as a political tool. He makes French hegemony over the bulk of North America starkly clear, and records the accomplishments of French explorers in numerous historical legends. The land demarcated as French even extends through the Carolina region briefly sought by Ribault and Laudonnière in the 1560s. The British would respond to this cartographic warfare by making maps which extended the boundaries of their diminutive Atlantic Seaboard colonies due west, over the mountains and across the continent.

California is mapped as an island. The origins of this celebrated error date back to Cortés and the medieval expectation of oceanic islands which he and his contemporaries harbored when they "discovered" lower California in the 1530s. Cortés, not unreasonably, believed, or wished to believe, that he had discovered an island.[297] Although Ulloa's follow-up mission indicated that the land was a continental peninsula, and although it was almost exclusively depicted as such into the first quarter of the seventeenth century, the question was never truly resolved until the mid-eighteenth century.

---

297. A vestige of this can be found in the "y. de la perle" of the Gastaldi/Ramusio map of America (page 100 in entry 29).

The seeds of the island-theory revival were planted by Francis Drake. When Drake attacked the Spanish-held west coast of South America, the Spaniards of Mexico, jittery about the English threat, feared that Drake had returned to England not by circumnavigating the globe, but rather by a new and fast route sailing through the California Sea and into a passage above North America. This was likely a morsel of "mis-information" deliberately "leaked" to Spanish authorities; Drake had apparently convinced an innocent Portuguese pilot, kidnapped to guide them through the Magellan Strait and subsequently released to Spanish officials, that such was their plan. Bogus "intelligence" alluding to a new route also appears to have been fed to Spanish informants in England. In any case, Drake's voyage clearly heightened Spanish anxiety. If in fact the British had discovered a quick and easy route into the South Sea, the entire Spanish empire along the Pacific coast of the New World was doomed. Their only salvation would be to find the passage themselves so that it could be fortified and secured before the British could do so. Thus, Spanish efforts to find a passage connecting the California Sea with the Strait of Anian intensified, and the "knowledge" that such a strait existed created a self-fullfilling prophecy : the strait was "discovered."

One Fray Antonio de la Ascension is credited with making the myth materialize. After accompanying the Vizcaino expedition of 1602-03 he wrote, in a personal account of the voyage, that the waters of the Gulf of California continue uninterrupted through the Strait of Anian, and as a result, beginning in 1622, mapmakers showed it thus for over a hundred years. In the words of Henry Briggs, an early champion of the island myth,

> *"California sometymes supposed to be a part of the western continent, but scince by a Spanish Charte taken by the Hollanders it is found to be a goodly island."* [298]

While de Fer was making this map, the demise of the island myth was already brewing. Some mariners, such as Juan Cavallero Carranco in 1668, had already suspected that it was part of the mainland; but the serious trend away from the island theory was the result of a missionary named Eusebio Francisco Kino. An Italian educated in Ingolstadt, in 1681 Kino went first to Baja California and then to Mexico. He arrived in the New World believing that California was a peninsula. Once there, however, he succumbed to the overwhelming local opinion to the contrary, now accepting it as an island. But eighteen years later, powerful circumstantial evidence swayed his mind once again, and he attempted to prove a land connection between Mexico and California by sending Indians to deliver letters to a Father Salvatierra in the newly-established Baja mission of Loreto; although none of the letters arrived, Kino maintained his belief that California was peninsular, and in 1701 he drew a map showing it as such. A manuscript of his map was sent to Paris from Madrid by the Jesuit Father Alcazar, from which an engraved version was produced which quickly began to influence some mapmakers, notably the French.[299] But Kino had never actually proven his claim by travelling from Sonora to the Pacific Coast, and his theory was not accepted by the Spanish Crown until 1746, when Father Ferdinando Consag led an expedition to the mouth of the Colorado River, sailing completely around the Gulf and settling the matter entirely.[300]

Vignettes illustrating various endeavors of the peoples and fauna of America fill the empty areas of de Fer's map, and panels containing historical and descriptive text, added about 1703, flank the sides. De Fer's famous beaver vignette reflects the importance of that animal to European-American destiny. The beaver had become an unwitting but paramount

---

298. From Briggs' map "The North Part of America," London, 1625. The first maps to revert back to an insular California were the maps on the titlepages of Herrera's Descriptio Indiae Occidentalis and Le Maire's Spieghel der Australische Navigatie, both published in Amsterdam in 1622.

299. Kino's influence even anticipated his map. The landmark in rejoining California to the mainland was de l'Isle's map of North America, 1700.

300. As a result of Consag's voyage, Ferdinand VII issued a decree the following year proclaiming that California was not an island.

player in economic affairs in Canada, as well as in Dutch and English New York, because hats made from their pelts were highly popular in Europe. The search for pelts increased exploration by luring traders far into unknown regions, helping to shape French-Indian relations, and provide a commercial basis for the country which became modern Canada. De Fer explains the beavers' various enterprises in a legend below the illustration. Beavers "A" are the butchers or loggers who cut and fell trees, forming the base for their roadway; beavers "B" cut the logs to correct size; beavers "C" transport the cut logs to the building site; beavers "D" prepare mortar; beaver "E" is the supervisor or architect; beaver "F" is a doctor who tends to beaver wounds; beavers "G" put mortar on their tails; "H" is a beaver on his back from overexertion; "I" is a mason; beavers "L" compact the mortar for strength; "M" is a beaver lodging with two doors, one facing the land and one facing the water. De Fer, following romanticized accounts of beaver society, errs in such details as showing the door of the lodge facing the land, in showing beavers walking upright, and in their carrying mud on their tails.[301] The scene of Niagara Falls in the background is copied from a view included in Hennepin's book of the previous year (1697), which was the first printed view of the Falls.

De Fer's illustrations, particularly those of beaver endeavors at Niagara Falls and of the codfishing industry off the Grand Banks, were copied by subsequent mapmakers, notably Hermann Moll (1715) and Henri Chatelain (1719).

This separately produced work of de Fer was re-issued, apparently in greater numbers, in 1705. The present example is possibly the only survivor of the original 1698 issue with the full text panels intact.

- - - - - - - - - - - -

301. See Edward Dahl, "The Original Beaver Map - De Fer's 1698 Wall Map of America" in The Map Collector, issue 29 (December 1984), p. 22-26.

Chapter VIII

THE DUTCH

49. NEW YORK / NEW ENGLAND

Nova Belgica et Anglia Nova.

["New Belgium and New England"]. [in *America, Quae Est Geographiae Blauiannae Pars Quinta* . . .] Willem Blaeu, Amsterdam, 1635 (Joan Blaeu, 1662).

Medium : copperplate engraving, with original hand color.
Size of original : 390 x 510 mm.

COLOR PLATE : XXI

At the turn of the seventeenth century Dutch entrepeneurs looked to establish profitable commercial outposts in the non-European world. The Asian market was secured for the Dutch East India Company by a charter of 1602, and others looked to North America as a possible new source of trade. Although lacking the venerated goods of the Orient, North America nonetheless offered compensations: the voyage there was shorter and safer, no Dutch monopoly had yet been established, and the hope still lingered that a new route into the Pacific would be found along its coasts. In 1609 the Dutch East India Company sent the Englishman Henry Hudson, a friend of John Smith, to explore the river which now bears his name. Hudson had already made two voyages under the British flag in search of a passage to the Orient. Although the Hudson River failed to yield such a passage, its valley proved to be both unfortified by other European powers and rich in fur animals, providing an opportunity to upset France's self-proclaimed monopoly on the fur trade in North America. To this end in 1613 a consort of Dutch fur traders sent a skipper by the name of Adriaen Block to the Hudson Valley.[302] Scouting the Northeast Coast in 1614, he drew a chart which provided much of the data upon which Blaeu drew to compose this map of New England.

Blaeu's map owes its general coastal geography to Block's chart, notably his delineation of Manhattan, which Block was the first to correctly chart as an island. Information was also derived from a map of Johannes de Laet (1630), and a map of Samuel de Champlain (1613), upon which Blaeu's orientation of Lake Champlain ultimately depends. Blaeu illustrates the map with American fauna fresh to European eyes : the beaver, the polecat, and the otter, eating fish. The map is oriented with west at the top.

Willem Janszoon Blaeu was a student of the astronomer Tycho Brahe. He and his son, Johannes, created a monumental atlas of the world, of which the present volume, relating entirely to America, first appeared in 1635. This enlarged issue of 1662 contains the original *Nova Belgica Et Anglia Nova* as well as new maps of other parts of America.

- - - - - - - - - - - - - -

---

302. Earlier voyages of Block in 1611 and 1612 are known from inference. See Condon, New York Beginnings, p. 15.

50. NEW YORK / NEW ENGLAND

Novi Belgii Novaeque Angliae Nec Non Partis Virginiae Tabula multis in locis emendata a Nicolao Joannis Visschero.

["New Belgium, New England, and also part of Virginia, compiled and corrected by Nicolas Johann Visscher"]. Nicolas Visscher, Amsterdam, 1655. State IV.

Medium : copperplate engraving, with original hand color.
Size of original : 470 x 555 mm.

COLOR PLATE : XXII

Although Dutch interest in a commercial base in North America began as early as did designs for the East India Company, a lack of both a focused strategy and of funds delayed a concerted bid for the American enterprise until 1618. Not until 1621 was a charter for the Dutch West India Company finally granted. Prior to 1618, two conflicting philosophies —— that of a company which would essentially be an arm of the State, in the fashion of Spain's colonial successes in the Caribbean, and that of the company being principally a private/commerical venture but with colonial merit, following the example of the Dutch East India Company in the Southwest Pacific —— diffused efforts at a charter. When the West India Company's charter was resolved in 1621, it reflected both opinions.

In the midst of all this political haggling, four private companies had pursued trade in the region of New York and the Hudson Valley. But these firms —— the van Tweenhuysen, Hans Claesz, Witsen, and Hoorn Companies —— fought bitterly among themselves in a climate of opportunistic anarchy, and, as a result, none of them flourished. To solve the bickering and produce a viable trading organization, the New Netherland Company was formed in 1614 with a three-year charter. The framework for the Dutch West India Company was now established.

In 1623 the land earlier scouted by such agents of Dutch businessmen as Adrien Block and Henry Hudson became the Dutch colony of New Netherland. However, the same ambivalence which characterized the drawing of the West India Company's charter now plagued the new and fragile colony : neither Holland nor the West India Company showed sufficient commitment to their infant colony's welfare. The inhabitants of New Netherland were treated as mere itinerant extensions of the Dutch West India Company's commercial arm rather than as settlers in their own right. Their needs were neglected, and the incessant threat of harrassment from the neighboring British colonies often ignored.

This disconcerting proximity of the English was made inescapably obvious in a map published by the Amsterdam mapmaker Jan Jansson in 1651. The map may have been based on a manuscript of Augustus Hermann, one of New Netherland's more talented residents. The Bohemian-born Hermann, a deft and successful merchant in New Amsterdam, had himself already proven his willingness to take risky political stands by becoming one of the "nine men" who in 1647 challanged the totalitarian rule of Peter Stuyvesant.[303]

Jansson's printed map, markedly superior to any other available of the region, became a prototype for subsequent mapmakers through the mid-eighteenth century. Its first adaptation was this work by Visscher. Visscher made some corrections and additions to the Jansson map, and inserted a view of the beleagered colony of New Amsterdam on Manhattan Island. This quaint panorama of the early Dutch settlement is the third known printed view of New York.

Fear of the British threat proved valid, and the status quo in New Amsterdam lasted less than a decade after Visscher's map. The city lay in a tract of land which was "given" to the Duke of York by his brother, King Charles II, in the Spring of 1664. By September of that

---

303. Stuyvesant made his autocratic intentions clear upon his arrival in New York in 1647, and although he appointed nine advisors, he dissolved the group when they asked the Dutch government for redress of their grievances.

year the Dutch had yielded to a British fleet which had been deployed to secure York's claim. Shortly afterwards, the settlement on Manhattan Island was described by a Englishman thus :

> *"New-York is setled upon the West-end of . . . Manahatans Island, so called by the Indians . . . New York is built most of Bricks and Stone, and covered with red and black Tile, and the Land being high, it gives at a distance a pleasing Aspect to the spectators. The Inhabitants consist most of the English and Dutch, and have a considerable Trade with the Indians, for Bevers, Otter, Racoon skins, with other Furrs; As also for Bear, Deer, and Elke skins; and are supplied with Venison and Fowl in the Winter, and Fish in the Summer by the Indians, which they buy at an easir rate; And having the Country round about them, they are continually furnished with all such provisions as is needed for the life of man . . . The greatest part of the Island is very full of Timber, as Oaks white and red, Walnut-trees, Chesnut-trees, which yield store of Mast for Swine . . . as also Maples, Cedars, Saxifrage, Beach, Birch, Holly, Hazel, with many sorts more."* [304]

The local herbs included *"Purslain, white Orage, Egrimony, Violets, Penniroyal, Alicampane, besides Saxaparilla very common, with many more,"* and, in the month of May, roses and other flowers created a landscape so beautiful and aromatic *"that you may behold Nature contending with Art."*

- - - - - - - - - - - - - - -

51. NEW YORK / NEW ENGLAND

Pascaarte van Nieu Nederlandt.

["Sea chart of New Netherland"]. [in: *Atlas marin, Ou Monde Maritime. Contenant une briefe Description de toute les costes coqnues de la Terre. Nouvellement mis en lumiere, par Arnold Colom . . . Amsterdam, A. Colom . . .*] Arnold Colom, Amsterdam, circa 1653 (circa 1658).

Medium : copperplate engraving, with original hand color.
Size of original : 545 x 635 mm.

COLOR PLATE : XXIII

By the latter part of the seventeenth century a thriving and fiercely competitive trade in printed sea charts developed in Amsterdam. Its output varied from charts created from new, and therefore not necessarily reliable, data, to conservative works based on charts already published. Although the latter tended to be more decorative in design, in all cases the production quality was of the highest standards. Some were purchased for on-board use on the seas, while others became works of art to inspire a gentleman in his library.

In about 1653 or 1654 the Amsterdam chart-maker Arnold Colom produced a tall atlas of sea charts, the first true Dutch sea atlas of the world. Its charts of the oceans used the same scale as the contemporary Spanish and Portuguese charts (1:14 mill.). Such charts had heretofore been available only as manuscript portolan charts or printed on separate sheets, whether paper or vellum. Although Arnold's father Jacob also published a sea-atlas, it was

---

304. Daniel Denton, <u>A Brief Description of New-York</u>, London, 1670.

Arnold's, rather than his father's, which appeared first. Despite the excellence of Arnold's charts, the atlas did not fare well commercialy.

Ironically, Arnold was a tenant of the highly successful mapmaker Nicolas Visscher (see entry 50). In 1663 Arnold was forced to surrender his copperplates to Visscher to defray overdue rent. As Arnold's plates were never used again, not even by his father Jacob for his subsequent sea-atlas, it is probable that Visscher melted down the plates and re-used the copper for his own maps. The atlas is here opened to his *Pascaarte van Nieu Nederlandt*, the first printed map of New Netherland.[305]

Colom's map extends from the Outer Banks at 35° through Cape Ann *(C. An)* on the New England coast. Cape Ann, *Salem*, Boston *(Baston)*, and Plymouth *(Pleymuyt)* are all shown in their correct positions. Cape Cod is *Staten Hoeck*, but the term *C. Cod* appears at the eastern tip of Nantucket *(Natocket)*. Martha's Vineyard is *Martin Viguter Ey.* and has a *C. Ack* on the southwest coast (Squibnocket Point?). Below it is *Hendrick Cristiaens Eylandt* (No Mans Land). Block Island is *Ad.ʳ Blocx Eylandt*, and Fishers Island is *Visschers I.* Along the Connecticut River are the regions of *Pequatoos, Sequins*, and *Nawaes*, with *Fort de Hoop* marked at about present-day Middletown. Long Island is *Langh Eylant alias Matowwacs* and its sound is *Siwanoys*. Within Long Island are the settlements of *Oester bay* (Oyster Bay) and *Crome Gou* (Corum). Manhattan is *Manhattaons*, with *Nieuw Amsterdam* marked on the mainland to the west. The Hudson River is shown through *Fort Orangie*, an early Dutch fur-trading post near present-day Albany, and *Mahikans*. The Delaware River is charted through *Fort Nassouaw*, a stronghold for the New Netherland Company, and upriver through what is now Trenton. *Schuy Kil* is the river on which Philadelphia would soon be built. The Chesapeake contains no coastal names except for the region of *Minquaes* at its northern end. At the entrance to the bay are *Smits Isle* (Smith Island), *C. Charel* (Cape Charles), and *Point Confor*. The British settlement of Jamestown is *Iens Toun*. The James River is prominent, and the Carolina coastal detail continues south through and beyond the *R. de Iordan* of Ayllón.[306]

- - - - - - - - - - - - - - -

---

305. But see entry 52 (Lootsman).

306. Ayllón named it in 1526 after one of his captains. It is thought to be either the Santee or Cape Fear River.

52. NEW YORK / NEW ENGLAND

Pascaerte van Nieu Nederlant, Virginiae, Nieu Engelant en Nova Francia van C. of Faire tot C Forchu / Amsterdam By Theunis Iacobsz op't water inde Lootsman.

["Sea chart of New Netherland, Virginia, New England and New France from Cape Faire to Cape Forchu."]. Theunis Jacobsz Lootsman, Amsterdam, 1666. [in: *L'Atlas de Mer, ou Monde Aquatique, remonstrant toutes les Cotes de la Mer, a scavoir de la partie connue de l'Univers, avec une generale et exacte description d'icelles. Nouvellement mis en lumiere. Amst., Jacques et Gaspar Anthoine, 1668*].

Medium : copperplate engraving, with original hand color.
Size of original : 460 x 585 mm.

COLOR PLATE : XXIV

Theunis Jacobsz added the adjective *Lootsman* to his name to avoid confusion with other Amsterdam printers with the same surname. Following his untimely death his widow ——and eventually two of his sons ——took over the business.

Lootsman's *Pascaerte van Niev Nederlant* is first known to have appeared in his sea atlas in 1666, though the map attributes the father, who died in 1650, with authorship. The geography of the map supports the theory that it was indeed prepared by the elder Lootsman, and that it therefore was made no later than 1650, possibly giving it priority over the Arnold Colom chart as being the first of New Netherland.

Lootsman's chart covers the largest area of New Netherland charts, extending from below the Outer Banks through Nova Scotia. His Carolina geography resembles that on the Jodocus Hondius *Virginiae Item et Floridae* of 1606, and his rounded shape for Cape Cod (*C. Kod alias Staten hoeck*) is similar to that found on the de Laet map of 1630. Long Island is fat and divided in two by a strait. The only river shown is the Hudson, which extends north beyond *F. Orange* to *Mahicans* and *Maquaas*.

- - - - - - - - - - - - - - -

53. NEW YORK / PENNSYLVANIA / NEW ENGLAND

Chart of the Sea Coasts of New Nether Land, Virginia, New England, and Penn-Sylvania, with the Citty of Philadelphia, from Boston to Cabo Karrick / Pas-caert van Nieu Neder Land, Virginia, Nieu Enge Land Als mede Penn-Silvania, met de Stad Philadelfia, van Baston tot Carrik by Hendrick Doncker inde Nieuwenbrugsteeg.

Hendrick Doncker, Amsterdam, 1688. From Doncker's *Nieuwe Groote Vermeerderde Zee-Atlas.*

Medium : copperplate engraving, with original hand color.
Size of original : 505 x 585 mm.

COLOR PLATE : XXV

This rare work by Hendrick Doncker is among the most unusual and ambitious of the Dutch charts produced in Amsterdam during this active period. It is entirely revised from a *New*

Netherland chart Doncker had published in 1660, and it now contains an inset plan of the metropolis of a new English colony just formed by an interesting religious rebel.

The city is Philadelphia, and the rebel is of course William Penn. Although Penn's interest in American politics can be demonstrated only as early as about 1675, clearly his experiences of the previous decade had sown the seeds for his wish to escape to a new life. In 1662 Penn was expelled from Oxford for his failure to accept prescribed religious dogmas, and in 1668 he was imprisoned for writing against the doctrine of the Trinity. It was after this incarceration that he obtained a charter from Charles II for the establishment of a colony in America. The settlement was to be a "holy experiment" in which the individual would maintain complete freedom of religious and political thought. It was founded in 1681, though not chartered until 1701.

Doncker's inset plan of Philadelphia is derived from a plan made in 1682 by the colony's surveyor general, Thomas Holme. The plan reflects Penn's concept of a spacious and well-planned city, with open spaces and gardens.

Doncker appears to have consulted English sources for much of his map, evident in his geography of New England, and of course his inclusion of the Philadelphia plan. Regressively, he charts Boston north of Cape Ann (unmarked), possibly following a late issue of Captain Smith's map. The use of Smith's map (or similar prototype) is also suggested by his naming of Cape Cod as *Cabo Cod, oft C. Iames* ("Cape James" having been Smith's designation), and by the appearance of *R. Charles* and *Nieuw Plymouth* (rather than just *Plymouth*). Long Island bears a vastly different shape, being now fat and in that respect reminiscent of John Seller's map of 1676. Despite the evidence of English prototype and the acknowledgement of the new British colony of Pennsylvania, Doncker's political persuasion is blatantly Dutch. His country's old colony of New Netherland, though long ceded to Britain, has been retained on a large scale. Its indulgent boundaries stretch from New Jersey north through much of southern New England, squeezing British claims to the slender Atlantic coast east of Narragansett Bay.

- - - - - - - - - - - - - - - -

Chapter IX

PRE- AND POST- REVOLUTION

54. COLONIAL NORTH AMERICA

A Map of the British Empire in America with the French and Spanish Settlements adjacent thereto.

[shown open to "key" map].
Henry Popple, London, 1733.

Medium : copperplate engraving.
Size of original : 510 x 495 mm (key map only).

PLATE : 33

The eighteenth century saw British-French territorial disputes in North America escalate to outright warfare. France, uncontested master of the Mississippi, claimed all land drained not only by that river, but by any of its tributaries as well. Her cartographers engineered maps which stressed that advantage, showing the great river and its numerous branches consuming much of the continent like a squid's tentacles grasping its prey. British possessions, however, remained confined to the Atlantic coast east of the Appalachians. Her colonists looked to gobble up Spanish Florida on the south, and fought the French over Nova Scotia on the north, both for the land itself and for the fish in its waters. Her seizing of New York from the Dutch in the later seventeenth century had only exacerbated these tensions, because French trappers travelled the northern part of that region.

Prior to the consumation of armed battle, however, a vicious war over these claims was fought via the printed map. The medium of printing, more than just a method of mass production, lent an air of legitimacy and officiality to boundaries which manuscript maps were incapable of doing. Thus the French printed maps of the seventeenth century had culminated with a clear statement that it was they who were winning the grab-bag to claim the North American interior (see de Fer, entry 48).

The eighteenth turned dirtier. The clearest example of the mapmaker's warfare came in 1718 when the French geographer Guillaume de l'Isle published a map which choked the British colonial border even further east than the Appalachian frontier. He also initiated bitter antagonism over the question of historical precedent in Carolina. Curiously, both France and England had laid claim to that region in the name of a "Charles". For France it was their strongholds at *Charlesfort* (1562) and *Fort Caroline* (1564), both established during the reign of the puppet/child King Charles IX. De l'Isle records this in a legend but twists it (in British eyes) to mean that *all* of Carolina was so-named by the French: *"Caroline ainsi Nommez en l'honneur de Charles 9 par les Francois qui le decouvrirent . . .".* In the seventeenth century the British dubbed their own Carolina colony, begun as early as 1629, in deference to *their* Charles I. In 1720, responding to de l'Isle, the English mapmaker Hermann Moll published a map criticizing these French claims; and so the bickering continued. But soon the English colonies realized that what they needed was not a map to carry yet more petty propaganda, but rather one which would provide the most accurate information obtainable about the entire colonial sphere of French and British claims, so that the inevitable days of reckoning to come could be planned for —and fought —intelligently.

Such was the genesis of Henry Popple's map. Popple, who was clerk to the Board of Trade and cashier to Queen Anne, set about creating a map which would be a practical tool for political ends, rather than a political tool in itself. Seeking the finest data available he therefore used, ironically, French maps for most of his delineations. These included, most insultingly to British pride, the de l'Isle *Carte de la Louisiane* of 1718 which had inflamed vanities over the issue of Carolina history, and the de Fer *La France Occidentale dans L'Amerique Septentrionale* of the same year. A British source, Colonel Barnwell's manuscript wall map of circa 1722 (or a copy of it), provided a basis for the Southeast mapping. Evidence of Cadwallader Colden's *A Map of the Countrey of the Five Nations* (1724) is found in Popple's delineation of New York. Popple included a wealth of detailed geographic information, particularly in the Great Lakes region, to which the British conducted trading expeditions from New York.

PLATE 33   Henry Popple, London, 1733   (entry 54)

The finished product, which was the first large-scale English mapping of North America, was published as an atlas of twenty sheets which could be fitted together if desired. It suffered two flaws, however, one a practical issue, the other an emotional one, and so the map received a mixed reception. As Popple's French sources were already a bit dated, some current French maps were in respects superior to Popple's. Also, in copying his French sources, Popple included many of their proposed boundaries as well, which though invaluable for planning strategy, was indiscreet and hardly appealing to much of the map's nationalistic English audience. Fort Niagara, for example, lies in French territory on Popple's map. Thus Popple was seen by many British subjects to have participated in the cartographic carnage of their adversary. [307]

- - - - - - - - - - - - - -

55. NEW ENGLAND

A correct map of the coast of New England.

[in: *The English Pilot - The Fourth Book.*] Mount & Page, London, 1755.

Medium : copperplate engraving.
Size of original : 480 x 635 mm.

PLATE : 34

The printed sea chart trade in England evolved out of the manuscript *Thames* or *Drapers'* school portolan tradition. In 1671 the instrument maker John Seller, a Baptist who had earlier aroused the wrath of the intolerant British royalty for his association with Nonconformists, received a privilege from King Charles II for a thirty-year monopoly for *The English Pilot* and *Sea Atlas*. These works were composed largely of re-worked Dutch plates; because of this element of commercial expediency, the temptation to make a pun on *sell* from Seller's name did not escape contemporary commentators. Within six years Seller was risking bankruptcy, however, and so he collaborated with two other chart-makers. One was William Fisher, whose publishing company was the pre-cursor to the Mount & Page firm; the other was John Thornton, a protégé of Nicholas Comberford's apprentice John Burston (see entry 43). Seller's project underwent a period of shaky financial arrangements and unstable partnerships, but the tradition of the English sea-atlas had finally been established.

In 1689 Thornton and Fisher created *The Fourth Book,* a new volume of *The English Pilot* which was exclusively devoted to American shores. This work was published throughout the eighteenth century by the firm of Mount & Page. Although the atlas was occasionally expanded and revised, the company's efforts to update its charts were lethargic and inadequate; the publisher's conservatism doubtlessly contributed to many pilots' troubles in American waters. *The Fourth Book* nonetheless remained the principal English sea-atlas of America during the eighteenth century.

307. As regards the de l'Isle Carolina controversy, it is however far from clear that it was not the British themselves who were toying with the coincidence of names when they named Carolina after Charles I. William Cumming and Helen Wallis, in their introduction to the Harry Margary facsimile of the Popple, judge the de l'Isle legend as having been "historically incorrect," but from the French viewpoint it might appear academic as to whether "Carolina" or "Caroline" was the name in question and whether it designated the entire region or just a fort. Legend quoted from de l'Isle's "Carte de la Louisiane" of 1718. See Cumming/Wallis for reference to Popple's geographic sources.

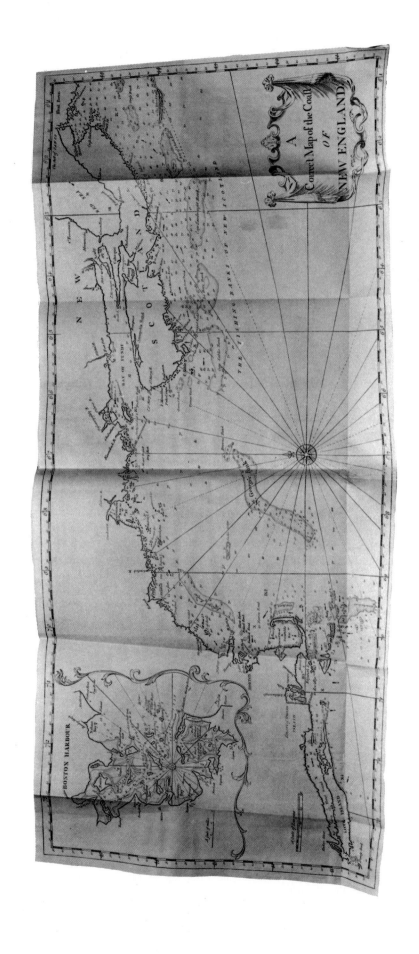

PLATE 34   Mount & Page, London, 1755   (entry 55)

56. COLONIAL NORTH AMERICA

A Map of the British and French Dominions in North America with the Roads, Distances, Limits, and Extent of the Settlements, Humbly Inscribed to the Right Honourable The Earl of Halifax, and the other Right Honourable The Lords Commissioners for Trade and Plantations . . .

John Mitchell, London, 1755. Separate publication.

Medium : copperplate engraving.
Size of original : eight sheets (unjoined), each 485 x 685 mm.

PLATE : 35, 36

John Mitchell was a physician interested in botany and zoology, and was posthumously famous for his treament of yellow fever. Historians differ as to when he was born, and whether he was born in Britain or Virginia, although he clearly lived many years in the latter. He might have been a Quaker, definitely befriended Benjamin Franklin, and strongly supported British, rather than French, sovereignty in America. In 1746 he went (or returned) to England, apparently abandoning both medicine and botany; the following year he was elected a Fellow of the Royal Society. Aside from this sketchy information, little is known about Dr. Mitchell except for one point : he made a map.

Work on his map began in 1750. In compiling it, Mitchell exhausted the many manuscript charts and records in the possession of both the British Admiralty and the British Board of Trade. After five years in the making, the map was issued under the auspices of the British government.

Mitchell's map records detailed geographical and topographical information, the location of Indian tribes, remarks about the nature of the country, and historical notes about European settlements. It became an immensely influential political tool, serving most significantly to establish boundaries in the Paris treaty negotiated in 1782-83 to end the Revolutionary War. It continued to be consulted as an authoritative document in numerous parleys through the nineteenth century; indeed, as late as a boundary dispute between New Jersey and Delaware in 1932.

The two sheets showing the Great Lakes, New York, and New England are illustrated.

- - - - - - - - - - - - -

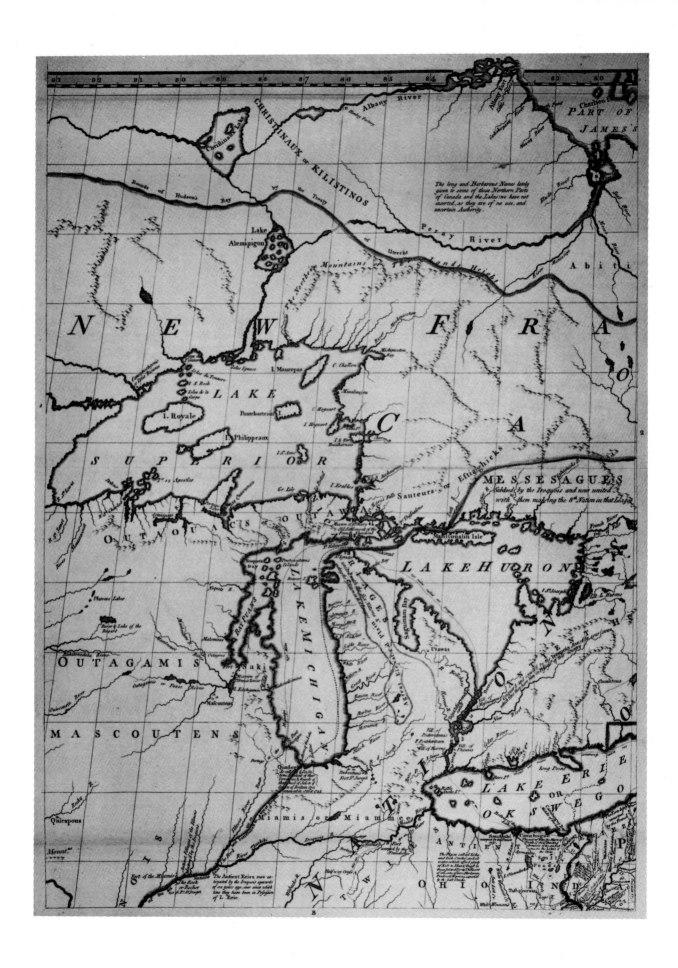

PLATE 35   John Mitchell, London, 1755   (entry 56)

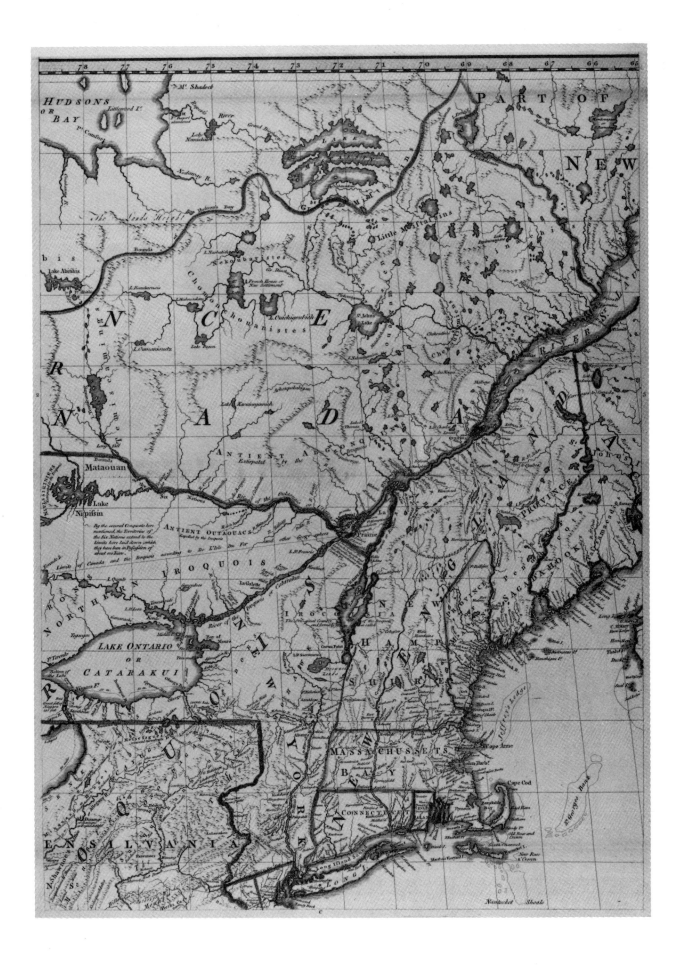

PLATE 36   John Mitchell, London, 1755   (entry 56)

57. MID-WEST, NEW ENGLAND

A Topographical Description of Such Parts of North America as are Contained in the (Annexed) Map of the Middle British Colonies (etc) . . . in North America. / London, Printed for J. Almon, opposite Burlington House, in Picadilly. MDCCLXXVI.

Lewis Evans, Philadelphia, 1755 /Thomas Pownall, London, 1776.

Medium : copperplate engraving.
Size of original : 510 x 865 mm.

PLATE : 37

Late in 1753, George Washington, then a 21 year old major in the colonial militia, was given the unpleasant task of delivering a letter of protest to the commander of French troops which had of late increasingly fortified the Ohio valley. French claims in the region seriously threatened the aspirations of the Ohio Company, which had been formed in Virginia three years earlier with a large grant of land from the King. But the French militia was predictably unimpressed by Washington's diplomacy, and so Virginia began to assemble an army to send into the Ohio region.

As a result, accurate mapping of these interior regions became more crucial than ever before. In response to this urgent demand, Lewis Evans, a native of Wales who had emigrated to the colonies and who had already established a fine reputation for mapmaking in resolving minor colonial disputes, was commissioned by the assembly of Pennsylvania to secretly compile a general map of the Middle Colonies.

The map covered the region from the Great Lakes south through the Carolinas. Two early drafts of the map were dispatched to General Braddock just prior to his unsuccessful bid against the French in 1755. The finished map was published in June of that year by his close friend Benjamin Franklin in Philadelphia.

The work was immediately recognized as a marked improvement in detail and accuracy over previous such mappings. Washington refers to it favorably in letters as early as 1756. It was in demand in both the colonies and in England, and was plagiarized by several publishers.

Evans' success, however, was quickly tempered by the alienation he provoked from many colonists over his outspoken political opinions. For one thing, he favored acknowledging French sovereignty over land northwest of the St. Lawrence from Fort Frontenac to Montreal, counter to both the conviction of Braddock's successor, General Shirley, and to the patriotic zeal then rousing the colonies. But more fatefully, he fell into a political rout with governor Morris of Pennsylvania over British colonial affairs, libeling the governor in the midst of it. He was jailed for the offense, and died in prison just 12 days short of the first anniversary of the publication of his map.

Following his death the plate for the map was acquired by Evans' dear friend and supporter, Thomas Pownall, and it is from this circumstance that the present map was born. Pownall had served in numerous capacities in colonial government affairs, including the governorships of Massachusetts Bay and South Carolina, and had been strongly supportive of the wars against the French. He advocated the unification of the colonies into a coherent dominion with tighter ties to England. In 1776 Pownall re-issued the map with text and geographic emendations; the proceeds were to go to help support Evans' daughter, Amelia, who had married an Irish ship captain employed in trade with the Levant, but like her father suffered from financial problems. However, owing to the popularity of plagiarized editions, demand for Pownall's issue was scant and only modest funds were raised.

In its accompanying text Pownall describes his changes to Evans' map, and strongly criticizes the various publishers who had copied the map in the intervening years.

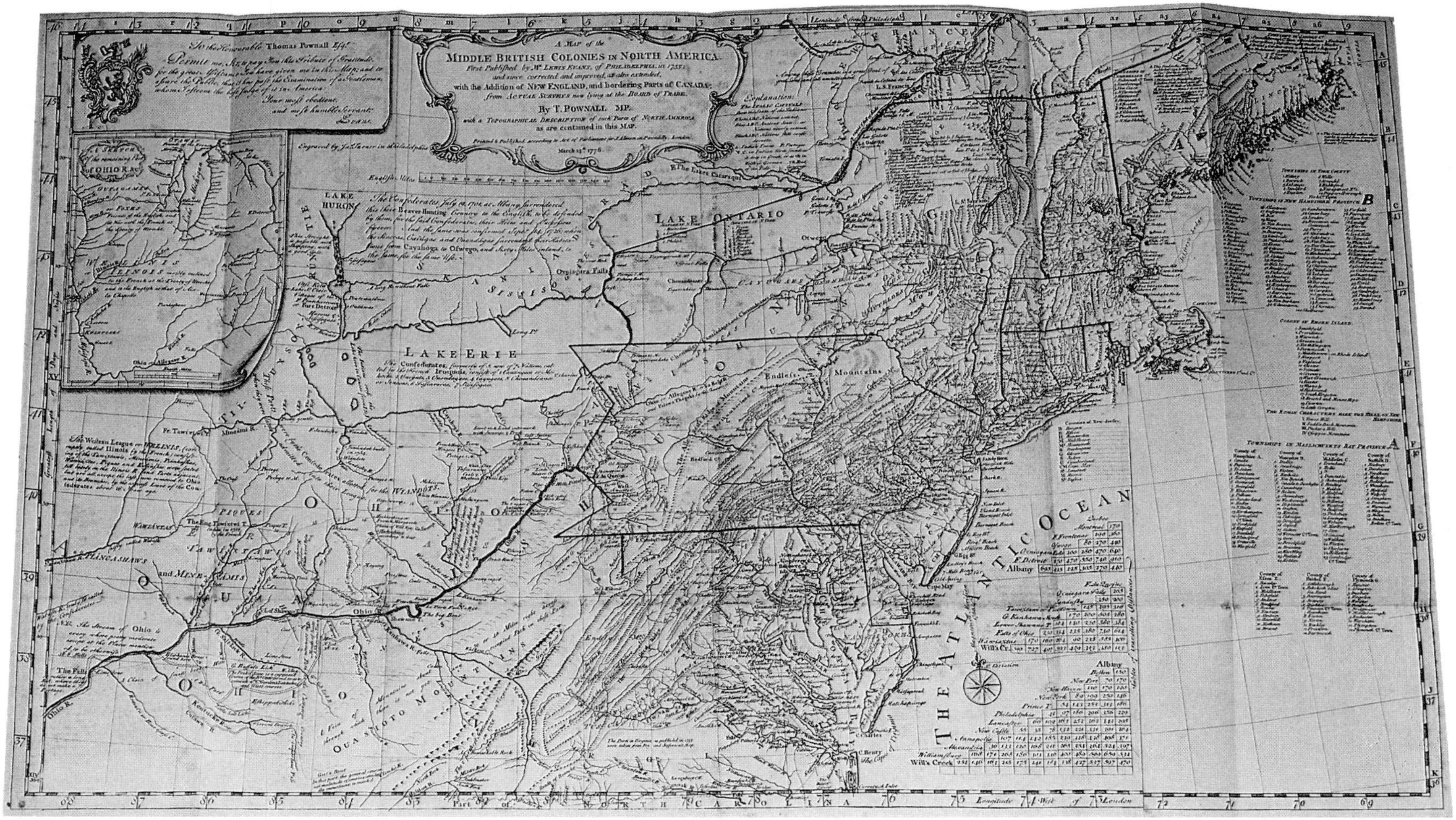

PLATE 37   Lewis Evans /Thomas Pownall, London, 1755 /1776   (entry 57)

Pownall has modified and added detail to the plate's New England region, and has also eliminated the original right-hand border of the map to facilitate the pasting of a new sheet which extends the map's coverage of the area by nearly 5 degrees. He has also indicated, with a dotted line, a variant theory as to the course of the Ohio River which had been suggested by the explorers Christopher Gist and Harry Gordon.

Evans' 1755 map contained the earliest known reference to oil in the region, the name *petroleum* appearing in two places: near what is now Oil City, Pennsylvania, and Wheeling, West Virginia.

- - - - - - - - - - - - - -

58. NEW ENGLAND

The Coast of Nova Scotia, New England, New York, Jersey, the Gulph and River of the St. Lawrence . . .

Joseph Frederick Wallet Des Barres, London, 1778.
From *The Atlantic Neptune*.

Medium : copperplate engraving.
Size of original : 1,195 x 840 mm.

PLATE : 38

In 1763 the Commander-in-Chief of the British forces in North America, Admiral Spry, told the British Admiralty that the available charts of the Northeast seaboard were unsatisfactory, and suggested that new surveys be undertaken. A Swiss-born engineer, Joseph des Barres, was chosen for the task. Des Barres had played an active role both in the siege of the French fort at Louisbourg under General Amherst in 1758, and in the siege of Quebec under Wolfe the following year.

Working intensively during the decade spanning 1764 to 1774, des Barres compiled surveys of the Atlantic Coast with painstaking care. Yet another decade was spent back in England preparing, engraving, and improving the charts. The published product appeared, ironically, just when the United States had been established and Britain's loss of the colonies was all but final. Publication took place over a period of several years, beginning in 1777, under the name *The Atlantic Neptune*.

The charts created new standards of accuracy and detail and attempted to employ standardized symbols to indicate various topographical features. In addition to conventional depth markings, some depths are marked in circles to *"denote that Lead did not strike Ground at that depth of Water."* "M" denotes mud, "S" is sand, "Sh" is shells, "Sl" is slate, "O" is *ouze* (ooze), "Wh:S" is white sand, "F:s" is fine sand, "G" is gravel, "St" is *stoney,* and "R" warns of rocks. For their level of refinement and attempt at convention they can be said to mark a precursor to modern cartography. Inasmuch as the previous statement is true, it interesting to note that among those who assisted des Barres in surveying was James Cook. After his assignment as a surveyor for the project, Cook set out on his three famous voyages around the world. These voyages did much to tie together bits of the world which had still escaped European scrutiny, successfully determined longitude by use of the newly-invented chronometer, and established a more scientific approach to exploration.

- - - - - - - - - - - - - -

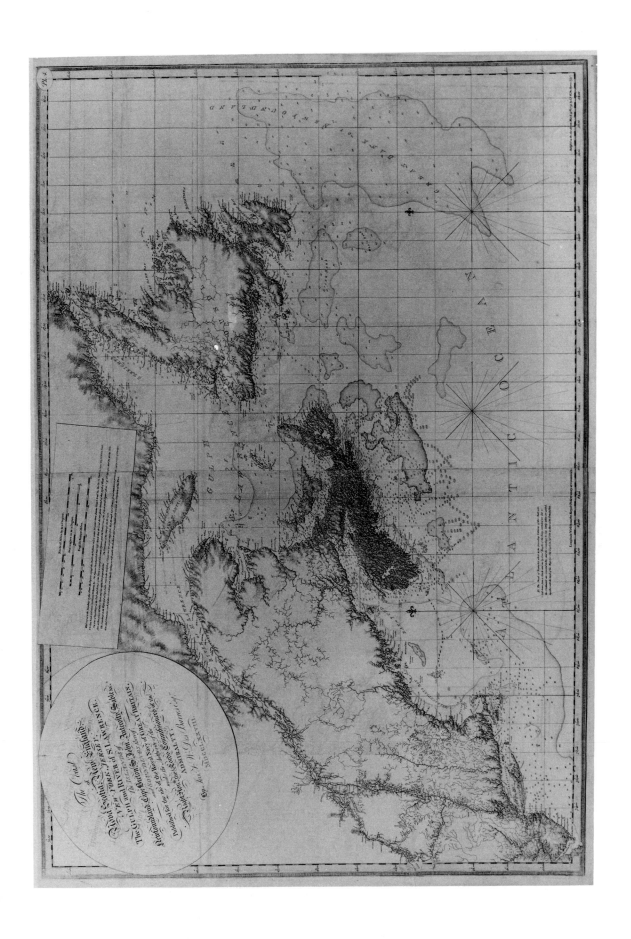

PLATE 38   Joseph des Barres, London, 1778   (entry 58)

59. NORTHEAST COAST OF NORTH AMERICA

A New Chart of America with the Harbors of New York, Boston, &c Drawn from the latest Authorities by W. Heather. 1799.

William Heather, London, 1799.

Medium : copperplate engraving.
Size of original : 635 x 787 mm.

PLATE : 39

In contrast to the utilitarian and lean design of the government-sponsored maps of des Barres, the private English maker William Heather produced charts which retained the rhumb lines of portolan chart inspiration and boasted more flamboyantly executed titles. Heather's career as a dealer in charts and nautical instruments precedes *The Atlantic Neptune,* his shop having been established in London in 1765 while des Barres had only begun his surveys. He published charts and pilot guides through his death in 1812.

This map's diagonal orientation allowed the most efficient use of space on the map sheet. Peculiarly, the map does not extend further south than eastern Long Island and his insets cover only Boston and Delaware Bay, despite the fact that in his title he specifies that the map includes the harbor of New York.

In the title Heather also claims to have compiled his map from the "latest authorities." But in the map he refers to Massachusetts as the "Province" of Massachusetts Bay, thus still treating New England as a British colony. In doing so he ignores not only the practical realities of the American Revolution, but even his own country's recognition of the United States as a sovereign country at the Treaty of Paris sixteen years earlier. Heather does, however, betray his use of sources dating from the Revolutionary period. His inset of Boston Bay, for example, marks *"Ruins of Charlestown"* where the 1775 battle of Bunker Hill was fought.

Heather's reluctance to cede his country's grasp on her North American colonies is also revealed in his use of the generic term "America" in the title, avoiding the use of the name of the new country. During this era the term *America,* coined nearly three centuries earlier to honor a landfall Vespucci was believed to have made somewhere along the coasts of Venezuela to Brazil, began its transition to a synonym for "United States." While today any continental part of the Western Hemisphere is still considered to be part of "America," the term by itself has become a misnomer to denote the United States exclusively, a transposition which would undoubtably have baffled Waldseemüller.

Notations useful to the sailor are engraved on the map. South of Martha's Vineyard an inscription informs the reader that *"At Holmes Hole in Marthas Vineyard the Flood Tide divides Eastward and Westward."* Comments in Cape Cod Bay warn of rising tides. Various types of shoals are noted, with reference made to sand *"with bits of Shells," "Fine black and white sand," "Coarse Sand black white and yellow,"* and *"Coarse brown sand."* North of the productive *"Crab Bank"* are *"Clay Banks"* and a *"Barren Bank."* Channels through the Cape's tricky waters are carefully plotted.

- - - - - - - - - - - - - -

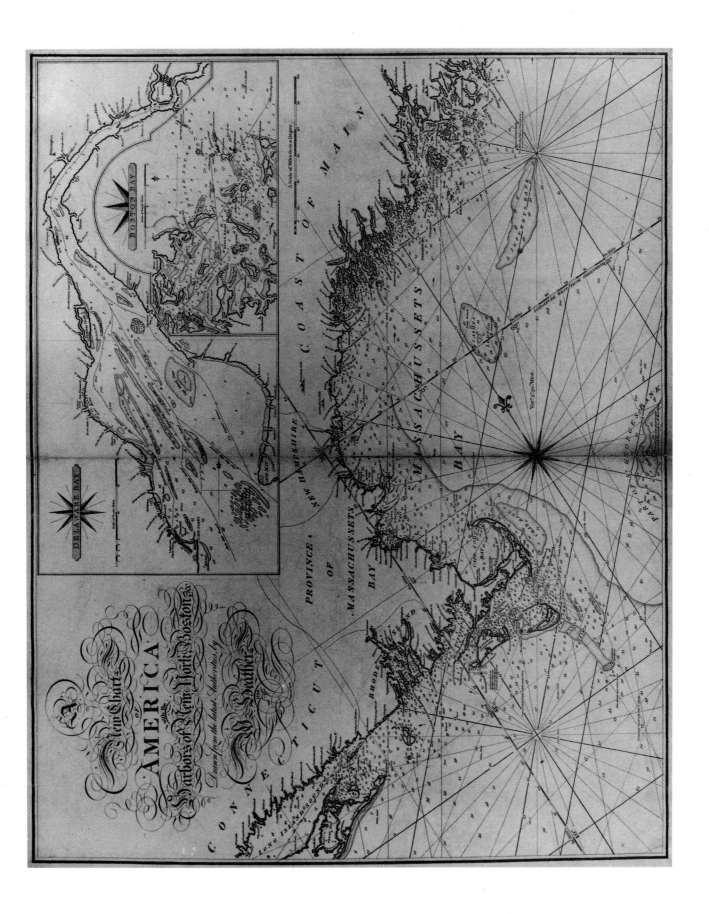

PLATE 39

William Heather
London, 1799

(entry 59)

60.   NEW ENGLAND

Chart from New York to Timber Island including Nantucket Shoals From the latest Surveys / I have carefully examined this chart and find it to agree with Hollands Surveys The Shoals by one well Authenticated by Bth Pilots. / Osgood Carleton Teacher of Navigation and other Branches of the Mathematics / Printed & Sold by W Norman at his Office n°-75 Newbury St. Boston.

[in: *The American Pilot,* William Norman, Boston, 1801.]

Medium : copperplate engraving.
Size of original : 990 (maximum) x 1375 mm.

PLATE : 40

Following the American Revolution, many New England sailors saw the need for improved and better detailed charts of the region's coasts. But in winning independence from Britain, the United States had also forfeited British interest in such projects, and neither the federal nor the state governments yet had the resources to focus attention on the matter. As a result, several enterprising people in the Boston area pooled their resources to produce charts locally.

Two of the most prominent figures in these efforts were Osgood Carleton and John Norman. Carleton was a surveyor, publisher of almanacs, and a highly respected teacher of astronomy and mathematics. He and Norman became associated as early as 1790, both in a publishing effort of their own and with that of Matthew Clark.[308]

Carleton's role was generally that of an endorser; the Boston Marine Society, which represented New England shipmasters, trusted him to judge the quality and accuracy of charts, his "seal of approval" (as seen in the title of this map) being an assurance to the purchaser of the map's integrity.

Private initiatives such as this satisfied Massachusetts' need for coastal charts through the end of the eighteenth century. The issue of an official state-sponsored map was addressed as early as 1792, however, and in 1797 the state formally drafted an agreement with Osgood Carleton and John Norman to produce such a map.[309]

John Norman passed to his son William both the chart trade and the collaboration with Carleton, producing charts such as this for the local New England market. Although these early American efforts generally lacked refinement, they nonetheless steered the republic through its difficult early period and showed great resourcefulness and cooperation on the part of the New England community. As will be seen in the next entry, private map publishing remained an important part of the American cartographic diet throughout the nineteenth century, despite the development of government surveys.

- - - - - - - - - - - - - -

308. Matthew Clark, a native of the village of Chilmark on Martha's Vineyard, published an atlas in 1790, at the age of 76. Clark offered for sale eighteen charts, available either separately or bound as an atlas; both Norman and Carleton were involved in this project.

309. In undertaking the project of an official state map, Massachusetts bore the additional responsibility of surveying Maine, a large and heretofore poorly mapped area over which Massachusetts had jurisdiction until Maine achieved statehood in 1820. See Susan Danforth, "The First Official Maps of Maine and Massachusetts," in Imago Mundi, vol. 35 (1983).

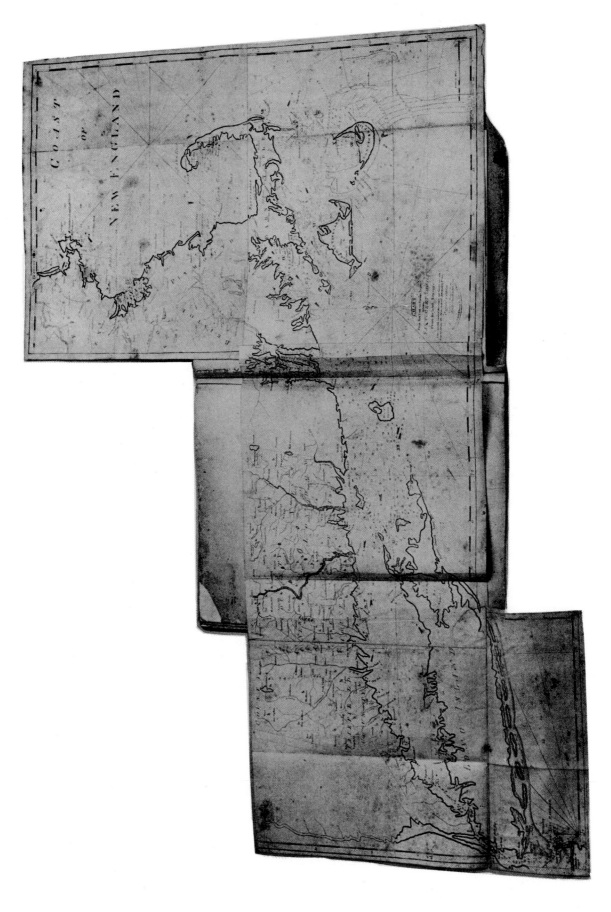

PLATE 40

William Norman
Boston, 1801

(entry 60)

61. MASSACHUSETTS COAST

Chart of the Vineyard Sound and Nantucket Shoals, Surveyed by George Eldridge, Hydrographer.

Published by S. Thaxter & Son, 125 State Street, Boston, 1865.

Medium : copperplate engraving.
Size of original : 1,0165 x 1,575 mm.

PLATE : 41

George Eldridge, like Carleton and Norman before him, applied Yankee deftness and private resourcefulness to the art of chartmaking. He was raised in Chatham, Massachusetts as a fisherman by his father, and obtained a ship of his own when still a young man. But his career at sea came to a dramatic end in 1851 when he was injured in a violent storm off the Massachusetts coast. Thus began the era of the Eldridge charts.

Most of the country's commerce was vulnerable to the notoriously treacherous waters off the New England coast. In an attempt to improve the safety of those seas, Eldridge combined coastal data gained through his own experience with that available from the U.S. Coast Survey charts (which were public domain), presenting the sum in a clear and easily readible, if inelegant, manner. He was able to update local detail much more efficiently than the government, and had a more intimate understanding of the format most useful to local sailors.

The largest publisher of charts in the United States at the mid-nineteenth century was the Blunt firm of New York. But for the typical New England sailor, neither those charts, concocted from U.S., French, and British government charts, nor the U.S. government charts themselves could compete with Eldridge's. Ironically, Eldridge used a relatively inexpensive lithographic process to print his charts, which was theoretically inferior to the electrotyping process used by the Coast Survey. But the government charts, if more handsome to look at, were less useful to a pilot under real-life conditions.

In 1886 a U.S. government commission established to study waste in its scientific agencies used the obvious preference that New England fishermen showed for the Eldridge charts as proof of government inefficiency. It concluded that Coast Survey charts had failed to meet the criterion of the maritime community, for

> *"had this been the case how could it have been possible for a private individual like Eldridge to adapt these charts to the needs of the seamen, and sell the reproductions at much higher prices than the originals."* [310]

By the early twentieth century the government switched to the lithographic process to print its charts, and adopted other hallmarks of the Eldridge charts, such as eliminating topographical information and making larger scales. Readability became an overriding concern. This, of course, negated Eldridge's advantage, and ultimately closed shop for the Eldridge charts. Eldridge's son published the firm's last chart in 1914. The pilot book survived, however ; the family has continued the annual publication to this day.

- - - - - - - - - - - - - - -

310. U.S. Congress, 1886. See Patrick McGlamery, "George Eldridge The Chatham Chartmaker," in Meridian, no. 3, 1990.

PLATE 41   George Eldridge, Boston, 1865   (entry 61)

INDEX

COLOR PLATES

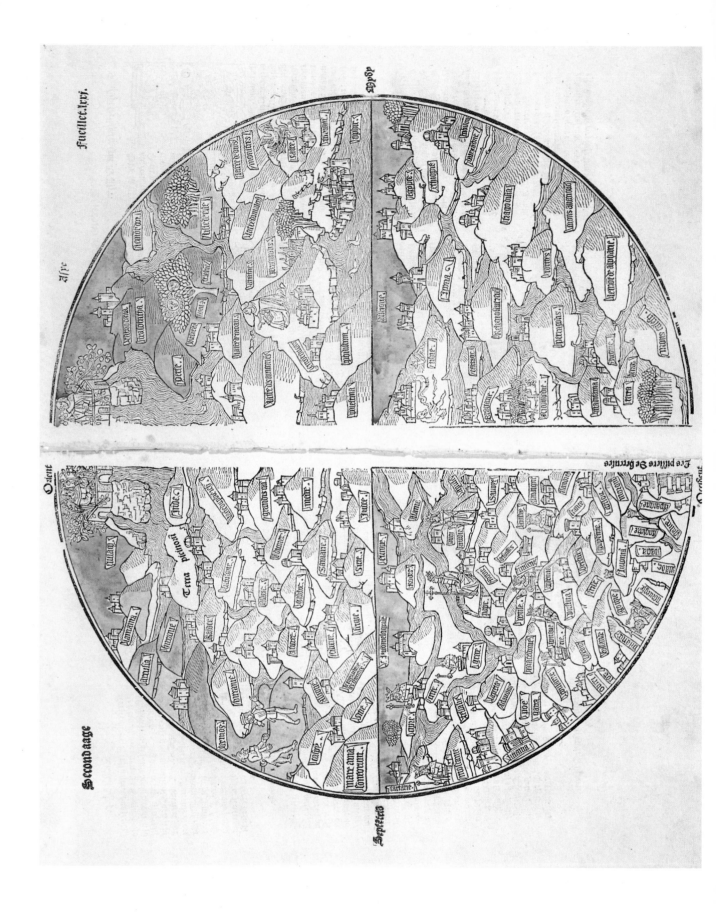

PLATE I

(anonymous)
Pierre le Rouge,
Paris, 1488

(entry 2)

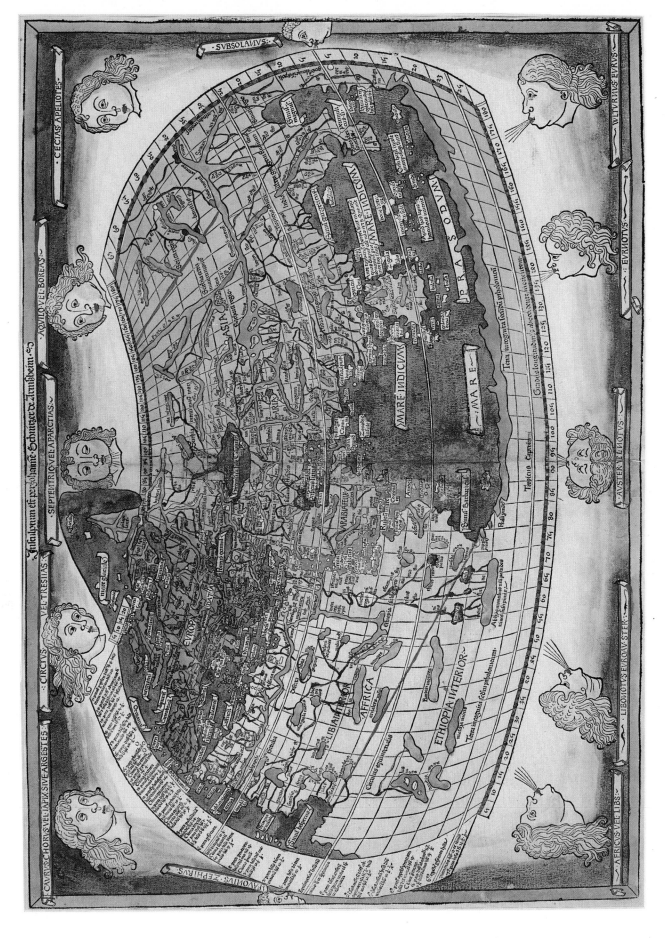

PLATE II

Claudius Ptolemy
Ulm, 1482

(entry 6)

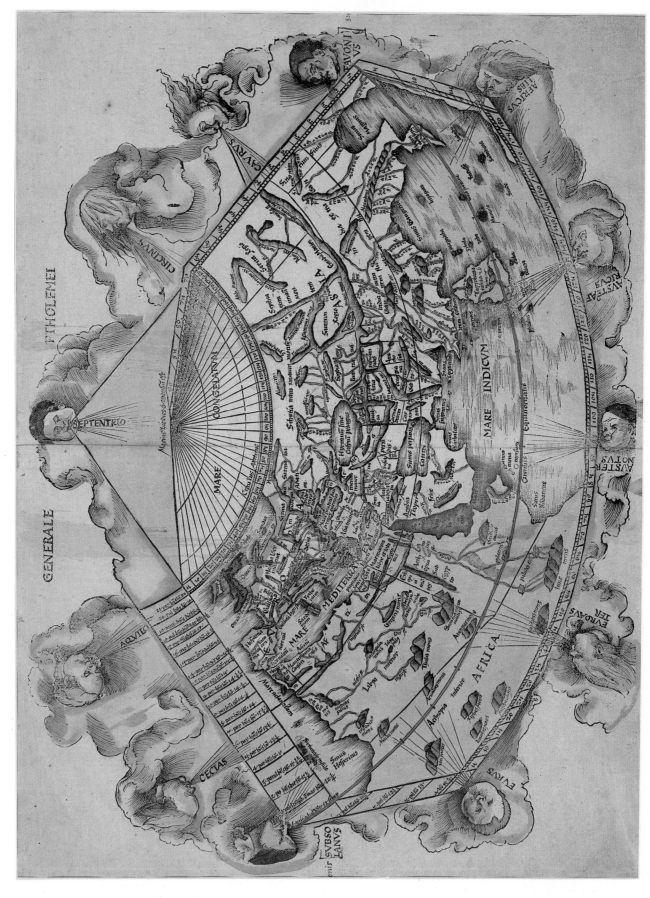

PLATE III

Claudius Ptolemy/
Martin Waldseemüller
Strassburg, 1513

(entry 8)

PLATE IV   Batista Beccari, Genoa, circa 1434   (entry 9)

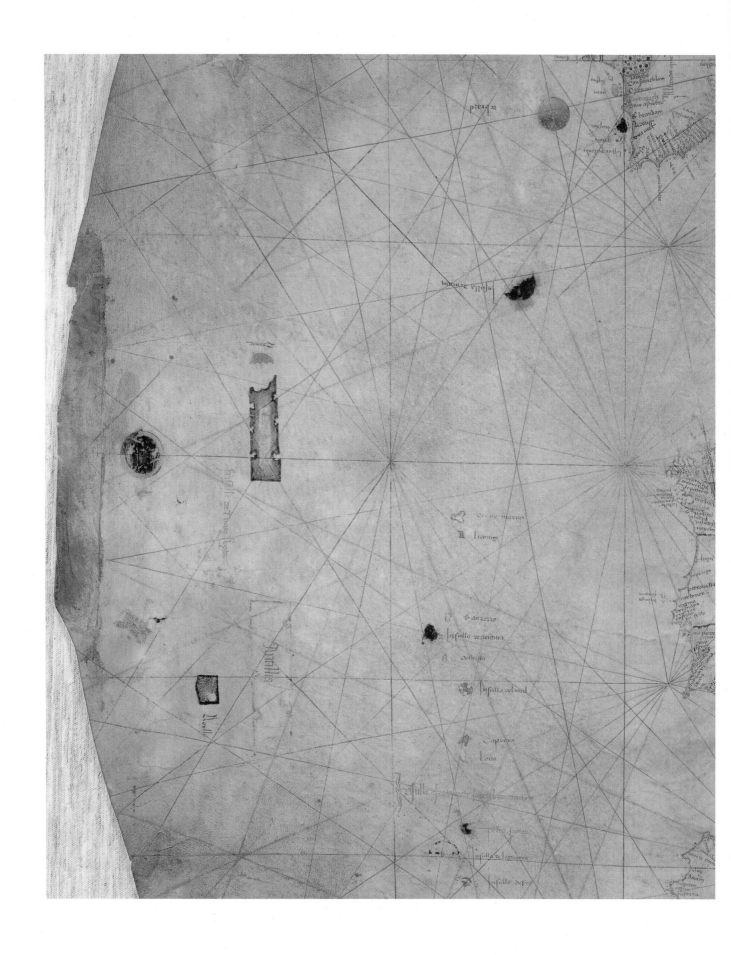

PLATE V   Batista Beccari, Genoa, circa 1434 (detail)   (entry 9)

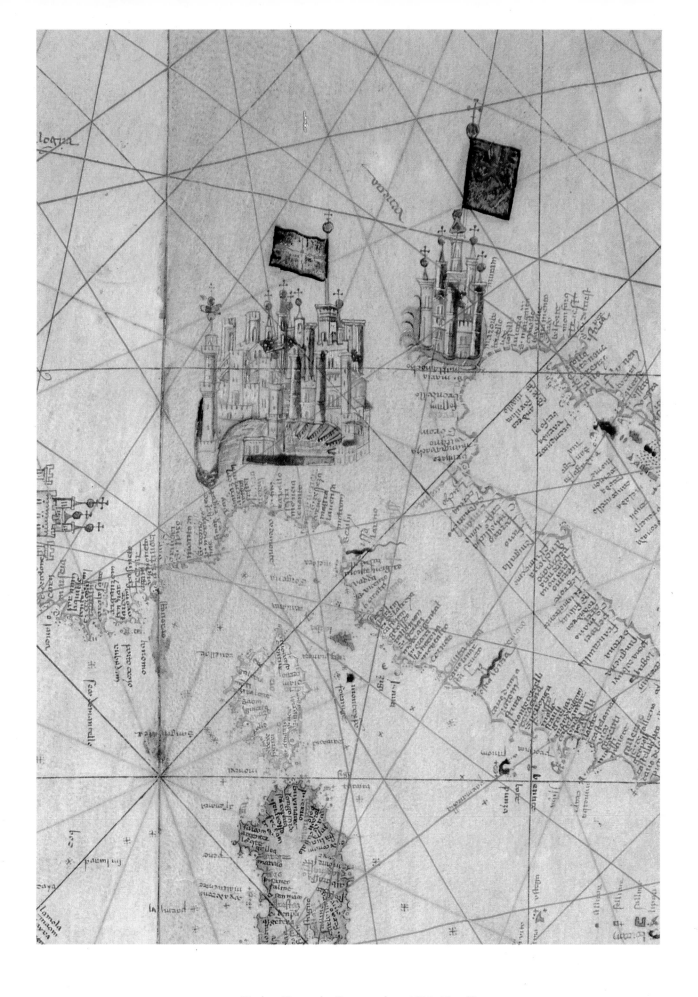

PLATE VI   Batista Beccari, Genoa, circa 1434 (detail)   (entry 9)

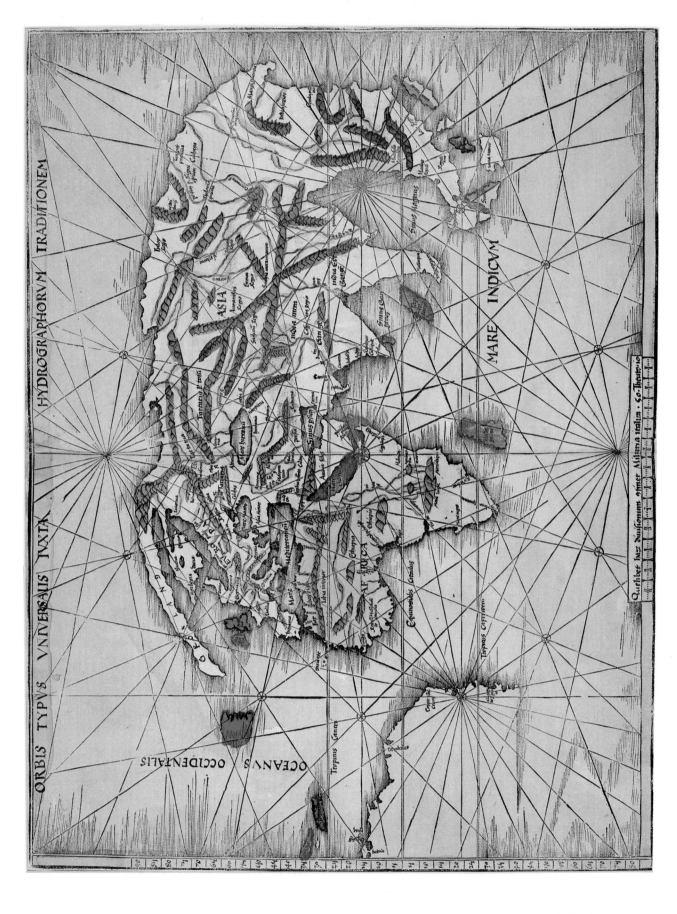

PLATE VII

Martin Waldseemüller
Strassburg,
1505-06 (?)

(entry 11)

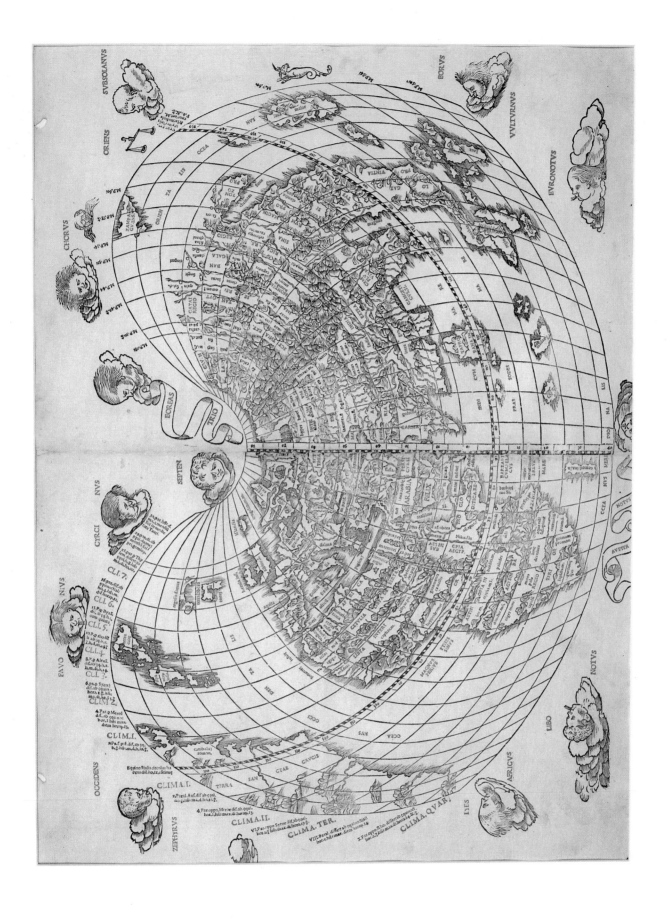

PLATE VIII

Bernard Sylvanus
Venice, 1511

(entry 14)

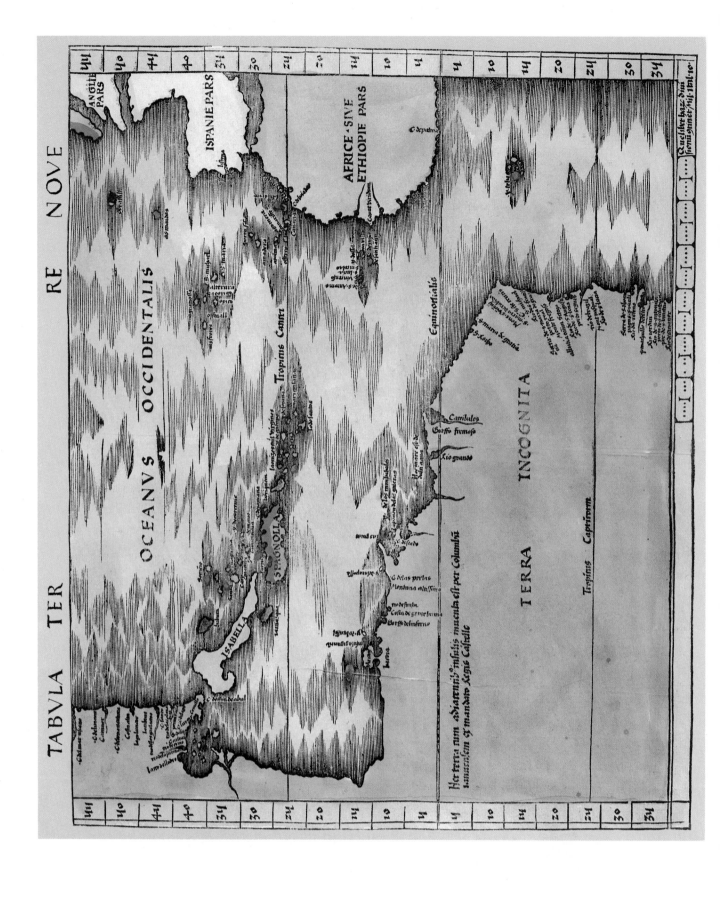

PLATE IX

Martin Waldseemüller
Strassburg, 1513

(entry 16)

PLATE X   Benedetto Bordone, Venice, 1528   (entry 18)

PLATE XI   Gerhard Mercator, Geneva, 1587   (entry 31)

PLATE XII   Mario Cartaro, Rome, 1577   (entry 33)

PLATE XIII   Mario Cartaro, Rome, 1577   (entry 33)

PLATE XIV   Lucas Janszoon Waghenaer, Leyden, 1583   (entry 34)

PLATE XV   Abraham Ortelius, Antwerp, 1589   (entry 35)

PLATE XVI

Willem Blaeu/
Pietro Todeschi
Amsterdam, 1608
(Bologna,
circa 1673)

(entry 36)

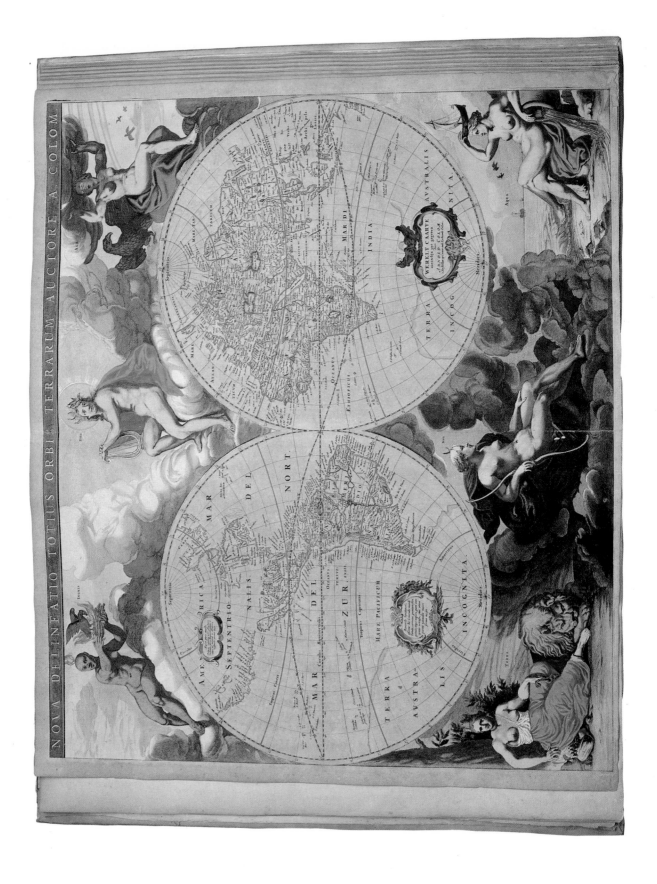

PLATE XVII

Arnold Colom
Amsterdam,
circa 1655

(entry 37)

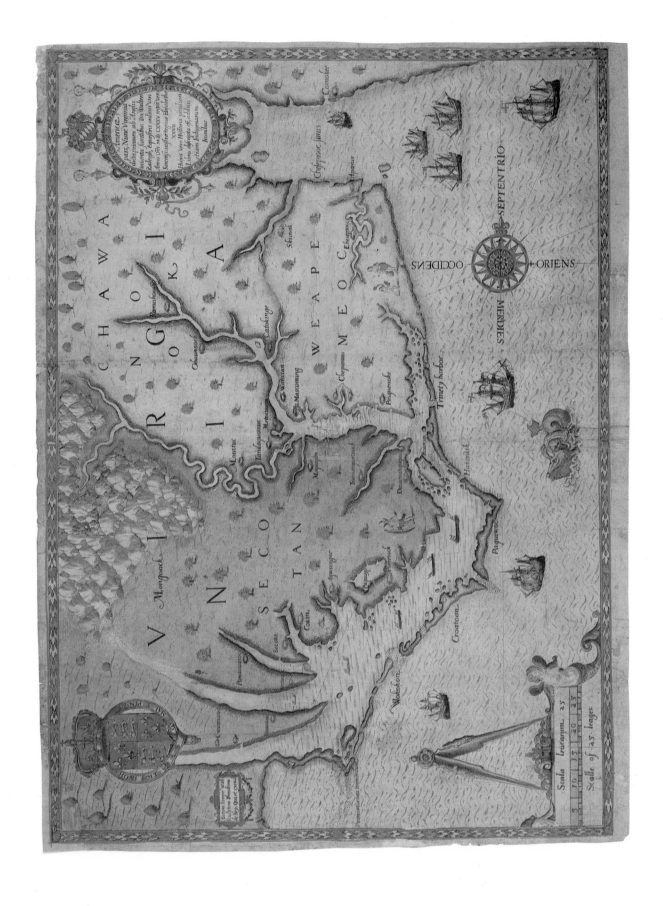

PLATE XVIII

John White/
Theodore de Bry
Frankfurt, 1590

(entry 39)

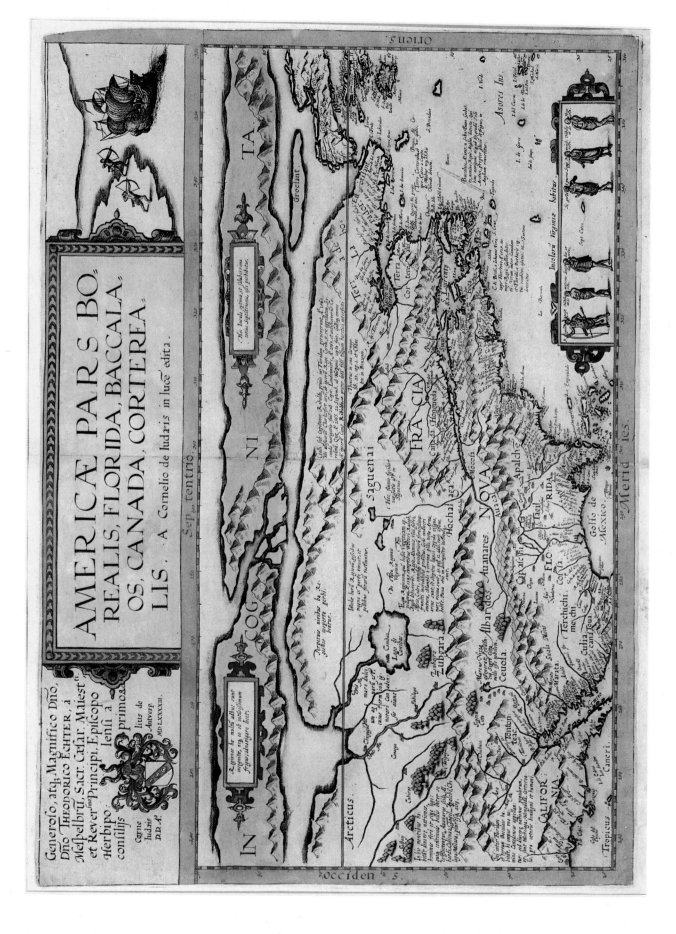

PLATE XIX   Cornelis de Jode, Antwerp, 1593   (entry 41)

PLATE XX   Nicholas Comberford, London, 1626   (entry 43)

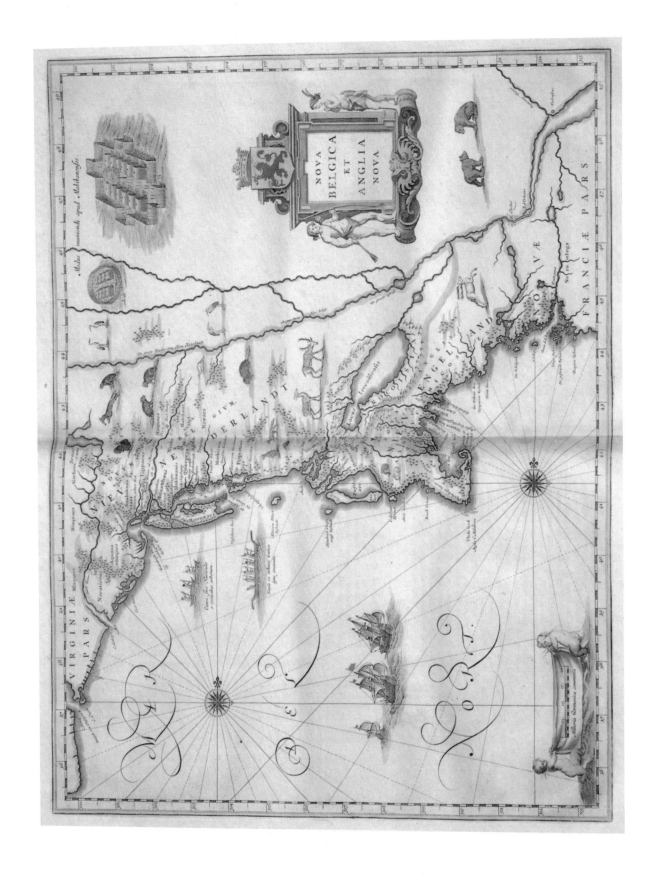

PLATE XXI

Willem Blaeu
Amsterdam, 1635

(entry 49)

PLATE XXII

Nicolas Visscher
Amsterdam, 1655

(entry 50)

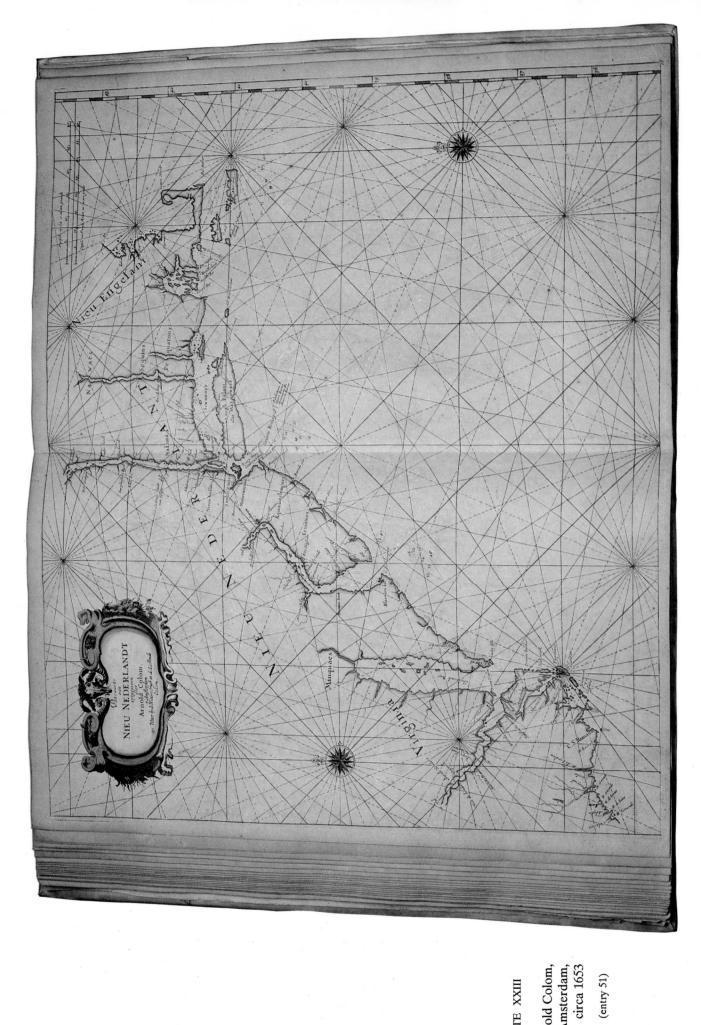

PLATE XXIII

Arnold Colom,
Amsterdam,
circa 1653

(entry 51)

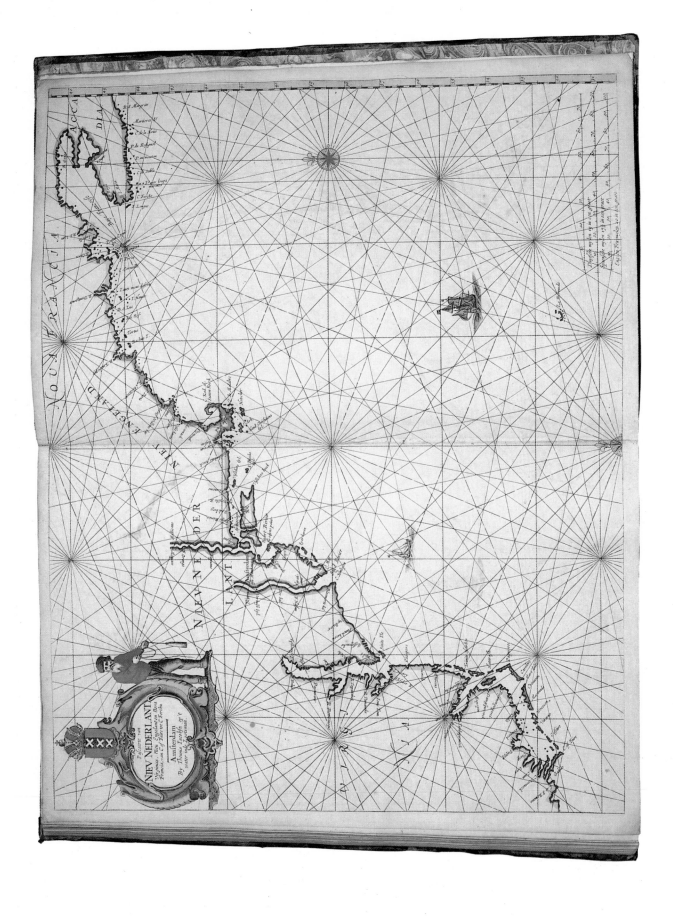

PLATE XXIV

Theunis Jacobsz
Lootsman
Amsterdam, 1666

(entry 52)

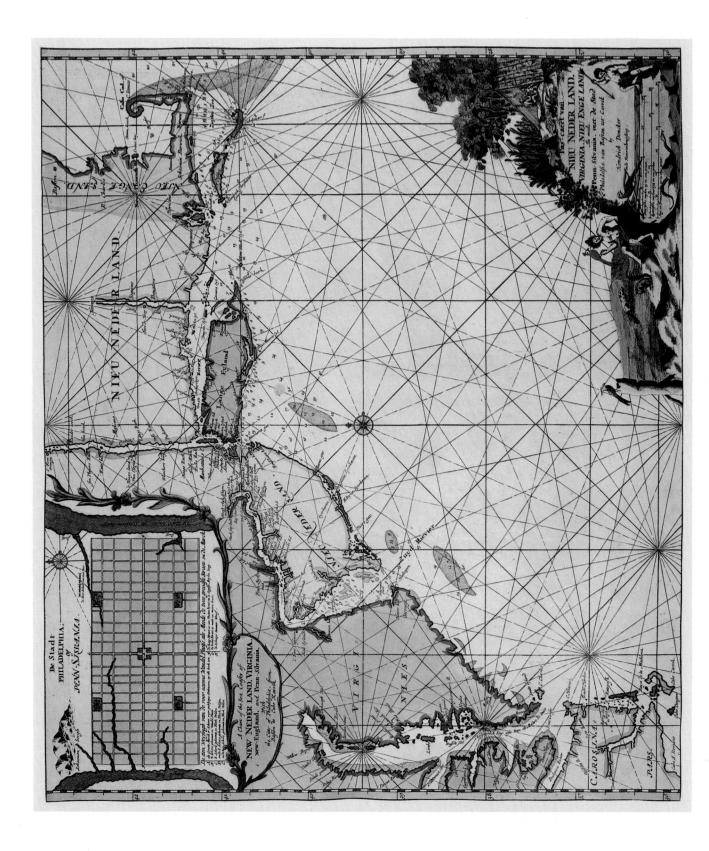

PLATE XXV

Henrick Doncker
Amsterdam, 1688

(entry 53)

BIBLIOGRAPHY

Alexander, Michael: *Discovering the New World.* (Harper & Row, 1976).

Anghiera: *(see Martyr, Peter)*

Bagrow, Leo: *History of Cartography.* (Precedent Publishing, Chicago, 1985, enlarged reprint of corrected 1966 edition).

Boxer, C.R.: *Women in Iberian Expansion Overseas 1415-1815.* (Oxford University Press, New York, 1975).

Brereton, John: *Discoverie of the North Part of Virginia.* (London, 1602; facsimile by Readex Microprint, 1966).

British Library: *Sir Francis Drake.* (British Museum Publications, 1977).

Brown, Lloyd: *The Story of Maps.* (Bonanza Books, New York, 1949).

Brown, Lloyd: *The World Encompassed.* (Walters Art Gallery, Baltimore, 1952).

Campbell, Tony: "The Drapers' Company and its school of seventeenth century chart-makers" in *My Head is a Map*, edited by Helen Wallis and Sarah Tyacke. (Francis Edwards, London, 1973).

Campbell, Tony: *The Earliest Printed Maps 1472-1500.* (The British Library, 1987).

Condon, Thomas: *New York Beginnings / The Commercial Origins of New Netherland.* (New York University Press, 1968).

Clissold, Stephen: *The Seven Cities of Cibola.* (Eyre & Spottiswoode, London, 1961).

Cumming, William: *British Maps of Colonial America.* (University of Chicago, 1974).

Cumming, William: *The Southeast in Early Maps.* (University of North Carolina, 1962).

Cumming, W.P. /Hillier, S.E. /Quinn, D.B. /Williams, G.: *The Exploration of North America.* (Paul Elek, London, 1974).

Cumming, W.P. /Skelton, R.A. /Quinn, D.B.: *The Discovery of North America.* (Paul Elek, London, 1971).

Cumming, W. /Wallis, H.: Introductory notes to facsimile of Henry Popple's *A Map of the British Empire* (Harry Margary, Kent, 1972).

Dahl, Edward: "The Original Beaver Map - De Fer's 1698 Wall Map of America." (in *The Map Collector,* December, 1984, p. 22-26).

Danforth, Susan: "The First Official Maps of Maine and Massachusetts." (in *Imago Mundi,* Vol. 35, 1983).

Denton, Daniel: *A Brief Description of New-York: Formerly Called New-Netherlands.* (London, 1670; facsimile by Readex Microprint, 1966).

Diringer, David: *The Book Before Printing.* (Dover, New York, 1982 {reprint of London, 1953 edition}).

Ethyl Corporation: *Lewis Evans and His Historic Map of 1755.* (Ethyl Corporation, New York {n.d.}).

Frabetti, Pietro: *Carte Nautiche Italiane dal XIV al XVII Secolo Conservate in Emilia-Romagna.* (Leo S. Olschki, Florence, 1978).

Fite, Emerson / Freeman, Archibald: *A Book of Old Maps Delineating American History from the Earliest Days Down to the Close of the Revolutionary War.* (Arno Press, New York, 1969 {reprint of Harvard University, Cambridge, 1926}).

Forbath, Peter: *The River Congo.* (Harper & Row, New York, 1977).

Gainer, Kim: "The Cartographic Evidence for the Columbus Landfall," (in *Terrae Incognitae,* vol. XX, 1988).

Gallo, Rodolfo: "Antonio Florian and his Mappemonde," (in *Imago Mundi,* vol. VI, 1949).

Ganong, W.F.: *Crucial Maps in the Early Cartography and Place-Nomenclature of the Atlantic Coast of Canada.* (University of Toronto Press, 1964).

Gómara, Francisco López de: *The Conquest of the Weast India.* (Thomas Nicholas' 1578 English translation of 1552 Spanish edition; facsimile by Readex Microprint, 1966).

Grattan, Hartley: *The Southwest Pacific to 1900.* (The University of Michigan Press, 1963).

Guthorn, Peter: *United States Coastal Charts 1783-1861.* (Schiffer Publishing, Exton, Pennsylvania, 1984).

Hapgood, Charles: *Map of the Ancient Sea Kings.* (E. P. Dutton, New York, revised edition 1979).

Hariot, Thomas: *A briefe and true report of the new founbdland of Virginia . . .* (Theodore de Bry, Frankfurt, 1590; facsimile by Readex Microprint, 1966).

Harley, J.B. / Woodward, David, eds.: *The History of Cartography. Volume I:*

Cartography in Prehistoric, Ancient, and Medieval Europe and the Mediterranean. (The University of Chicago Press, 1987).

Harrisse, Henry: *The Discovery of North America.* (N. Israel, Amsterdam, 1962 {reprint of London-Paris edition, 1892}).

Hemming, John: *Red Gold / The Conquest of the Brazilian Indians, 1500-1760.* (Harvard University Press, 1978).

Hough, Samuel: *The Italians and the Creation of America.* (John Carter Brown Library, Brown University, 1980).

Judge, Joseph, and Stanfield, James: "The Island of Landfall" (in *The National Geographic Magazine,* November, 1986).

Kelley, Jr., James: "Non-Mediterranean Influences that shaped the Atlantic in the early Portolan Charts," (in *Imago Mundi,* vol. 31, 1979).

Kelley, Jr., James: "The Map of the Bahamas implied by Chaves's 'Derrotero.' What is its relevance to the first landfall question?" (in *Imago Mundi,* vol. 42, 1990).

Koeman, C.: *Atlantes Neerlandici.* (five volumes, Theatrum Orbis Terrarvm, Amsterdam, 1967-71).

Kraus, H.P.: *Twenty-Five Manuscripts* (Catalogue Ninety-Five). (H.P. Kraus, New York {n.d.}).

Lach, Donald: *China in the Eyes of Europe.* (University of Chicago Press, 1965).

Lach, Donald: *Japan in the Eyes of Europe.* (University of Chicago Press, 1965).

Lanman, Jonathan: *Glimpses of History from Old Maps.* (Map Collector Publications Ltd., England, 1989).

Leithauser, Joachim: *Worlds Beyond the Horizon.* (Alfred A. Knopf, New York, 1955).

Letts, Malcolm: *Sir John Mandeville: The Man and his Book.* (The Batchworth Press, London, 1949).

Lessa, William: *Drake's Island of Thieves.* (University Press of Hawaii, 1975).

Martyr, Peter: *The Decades of the Newe Worlde or West India.* (Richard Eden's translation of Martyr's *De rebus oceanicis et orbe novo decades tres.* (Basle 1533, published London, William Powell, 1555; facsimile by Readex Microprint, 1966).

McGuirk, Donald: "Cuba on the Ruysch Map." (in *The Map Collector,* Issue 36, 1986).

McGuirk, Donald: "Ruysch World Map: Census and Commentary." (in *Imago Mundi,* Volume 41, 1989).

McGuirk, Donald: "The New World in an Old Cage." in *(IMCoS Journal,* vol. 4, no. 3).

Medina, Pedro de: *Libro de Cosmographia* (1538). (Translation by Ursula Lamb. The University of Chicago Press, 1972).

Morison, Samuel: *Admiral of the Ocean Sea.* (2 vols., Little, Brown, Boston, 1942).

Morison, Samuel: *The European Discovery of America / The Northern Voyages A.D. 500-1600.* (Oxford University Press, New York, 1971).

Mossiker, Frances: *Pocahontas.* (Alfred A Knopf, New York, 1976).

Muldoon, James: *Popes, Lawyers, and Infidels.* (University of Pennsylvania Press, 1979).

Münster, Sebastian: *A Treatyse of the Newe India.* (Richard Eden's London, 1553 translation of part of the fifth book of Münster's *Cosmographiae universalis.* (Basle, 1550; facsimile by Readex Microprint, 1966).

Nordenskiöld, A.E.: *Facsimile Atlas . . .* (Stockholm, 1889 {reprint by Dover, 1973}).

Nordenskiöld, A.E.: *Periplus.* (Stockholm, 1897 {translated reprint by Burt Franklin, Resource and Source Works Series no. 52, New York, n.d.}).

Ortelius, Abraham: *The Theatre of the World World.* (London, 1606; facsimile by Theatrvm Orbis Terrarvm Ltd., Amsterdam, 1968).

Osley, A. S.: *Mercator / A monograph . . . and translation of Ghim's Vita Mercatoris.* (Faber and Faber, London, 1969).

Parks, George: *Richard Hakluyt and the English Voyages.* (American Geographical Society, New York, 1928).

Parry, J.H.: *The Discovery of South America.* (Taplinger Publishing Company, New York, 1979).

Polk, Dora: *The Island of California.* (Arthur C. Clark Company, Spokane, Washington, 1991).

Polo, Marco: *The Travels.* (Translation by Ronald Latham. Penguin, Middlesex, 1958).

Pratt, Dallas: "Angel-Motors" (in *Columbia Library Columns,* Friends of Columbia Library, New York, May, 1972).

Purchas, Samuel: *His Pilgrimes.* (London, 1625).

Quirino, Carlos: *Philippine Cartography.* (Nico Israel, Amsterdam, 1963).

Revelli, Paolo: *Cristoforo Colombo e la Scuola Cartografica Genovese.* (Consiglio Nazionale delle Ricerche, Genoa, 1937).

Richardson, William: "The origin of place-names on maps," *The Map Collector,* issue 55, Summer, 1991.

Ristow, Walter, et al: *A la Carte.* (Library of Congress, Washington, 1972).

Sale, Kirkpatrick: *The Conquest of Paradise.* (Alfred A. Knopf, New York, 1990).

Sauer, Carl: *Northern Mists.* (Turtle Island Foundation, San Francisco, 1968).

Sauer, Carl: *Sixteenth Century North America.* (University of California, 1971).

Sauer, Carl: *The Early Spanish Main.* (University of California, 1966).

Scammell, G.V.: *The World Encompassed / the first European maritime empires c.800-1650.* (University of California Press, 1981).

Shirley, Rodney: *The Mapping of the World.* (Holland Press, London, 1984).

Skelton, R. A.: *Explorers' Maps.* (Spring Books, London, 1958).

Skelton, R. A.: Introduction to facsimile of Ptolemy's *Cosmographia,* Rome, 1478. (Theatrum Orbis Terrarvm Ltd., Amsterdam, 1966).

Skelton, R. A., et al: *The Vinland Map and the Tartar Relation.* (Yale University Press, 1965).

Smith, T. R.: "An Early Portolan of the Mediterranean by Nicolas Comberford, 1626." (in *Imago Mundi* vol. 29, 1977).

Spekke, Arnolds: *The Baltic Sea in Ancient Maps.* (M. Goppers, Stockholm, 1961).

Stahl, William: *Macrobius' Commentary on the Dream of Scipio.* (Columbia University Press, New York, 1952).

Stevens, Henry: *Ptolemy's Geography.* (The British Museum, 1908 {Orbis Terrarvm, Amsterdam, reprint}).

Stevenson, Edward: *Terrestrial and Celestial Globes, Their History and Construction.* (Yale University Press, New Haven, 1921).

Thacher, John: *The Continent of America.* (New York, 1896 {reprint Meridian Publishing, Amsterdam, 1971}).

Thrower, Norman (editor): *The Compleat Plattmaker.* (University of California Press, 1978).

Tilton, David: "Yucatan on the Peter Martyr Map?," (in *Terrae Incognitae,* vol. XXI, 1989).

Tooley, R. V.: *The Mapping of America.* (Holland Press Limited, London, 1980 {second impression, 1985}).

Verner, Coolie / Stuart-Stubbs, Basil: *The Northpart of America.* Academic Press Canada Limited, 1979.

Vorsey, Louis de: "Amerindian contributions to the mapping of North America: A preliminary view." (in *Imago Mundi,* vol. 30, 1978).

Wagner, Henry: *The Cartography of the Northwest Coast of America to the Year 1800.* (Berkeley, 1937 {reprint Nico Israel, Amsterdam, 1968}).

Waldseemüller, Martin: *Cosmographia Introductio,* 1507 (facsimile by Readex Microprint, 1966).

Wallis, Helen: *Raleigh & Roanoke.* (British Library Board and North Carolina Museum of History, 1985).

Welu, James: "The map in Vermeer's *Art of Painting*" (in *Imago Mundi,* vol. 30, 1978).

Wheat, J.C., & Brun, C.F.: *Maps & Charts Published in America Before 1800.* (Second revised edition, Holland Press Cartographica, 1978).

Winsor, Justin: *Narrative and Critical History of America.* (Eight volumes, Houghton, Mifflin and Company, Boston and New York /The Riverside Press, Cambridge, 1889).

Winsor, Justin: *A Bibliography of Ptolemy's Geography.* (Cambridge, 1884).

Winter, Heinrich: "Notes on the Worldmap in 'Rudimentum Novitorum'." (in *Imago Mundi* IX, 1952).

Woodward, David, et al: *Five Centuries of Map Printing.* (University of Chicago Press, 1975).

Woodward, David: "The Foster woodcut map controversy: a further examination of the evidence." (in *Imago Mundi* XXI, 1967).

Woodward, David: *The Maps and Prints of Paolo Forlani.* (The Newberry Library, Chicago, 1990).

Wroth, Lawrence: *Pacific Cartography.* (The Papers of the Bibliographical Society of America, Volume Thirty-Eight Number Two, 1944).

- - - - - - - - - - - - - -